Gait Analysis: An Introduction

Gait Analysis: An Introduction

Michael W. Whittle
BSc, MB, BS, MSc, PhD

Cline Chair of Rehabilitation Technology
The University of Tennessee at Chattanooga

formerly

Acting Director
Oxford Orthopaedic Engineering Centre
University of Oxford

BUTTERWORTH
HEINEMANN

Butterworth-Heinemann Ltd
Linacre House, Jordan Hill, Oxford OX2 8DP

A member of the Reed Elsevier group

OXFORD LONDON BOSTON
MUNICH NEW DELHI SINGAPORE SYDNEY
TOKYO TORONTO WELLINGTON

First published 1991
Reprinted 1993 (twice)

British Library Cataloguing in Publication Data
Whittle, Michael
  Gait analysis.
  1. Man. Movement
  I. Title
  612.76

ISBN 0 7506 0045 4

Cover illustration taken from Saleh M., Murdoch G. (1985). In defence of gait
analysis, *J. Bone Joint Surg.*, 67B, 237–41. With kind permission of author and
publisher.

Printed by Ipswich Book Company

Contents

Acknowledgements

First and foremost, I would like to thank my wife Wendy, and my four children James, Sally, Robert and Tracey, for their encouragement, and for (mostly) giving me peace and quiet while I was writing!

I would particularly like to thank three members of my former department at the University of Oxford. Derek Harris persuaded me that I should write such a book, and Jim Collins and Ros Jefferson reviewed my draft manuscripts, and made many helpful suggestions.

Thanks are also due to the very many professional colleagues whose ideas and opinions have contributed to this book, and particularly to Jim Walton, President of 4D Video, for information on some of the less well known kinematic systems. Figures 1.17–1.20 and 1.22–1.23 are reproduced from my chapter in 'Orthopaedics' (1987) by Hughes S P F, Benson M K D'A and Colton C L (eds.), with the kind permission of Churchill Livingstone.

Finally, I would like to thank Caroline Creed and the staff of Butterworth-Heinemann, who helped me to convert the idea into a reality.

This book is dedicated to:
Wendy, James, Sally, Robert and Tracey

Preface

Gait analysis is the systematic study of human walking. It is often helpful in the medical management of those diseases which affect the locomotor system. Over the past few years, there has been an increasing interest in the subject, particularly among practitioners and students of physical therapy, bioengineering, and several branches of medicine, including orthopedics, rheumatology, neurology and rehabilitation. Most previous books on the subject have been written for specialists, and are thus unsuitable for the student or general reader, since they assumed a certain amount of previous knowledge of the subject. I have attempted to write an introductory textbook, with the aim of providing the reader with a solid grounding in the subject, but without assuming a particular background or level of prior knowledge.

Chapter 1 is devoted to the basic sciences underlying gait analysis – anatomy, physiology and biomechanics. It is intended to give the reader who is new to these subjects the minimum required to make sense of gait analysis. It should also provide a refresher course for those who have once had such knowledge but forgotten it, as well as being a convenient source of reference material. Chapters 2 and 3 deal with normal and pathological gait respectively, showing the remarkable efficiency of the normal walking process and the various ways in which it may be affected by disease. Chapter 4 is devoted to methods of measurement, pointing out that gait analysis does not have to be difficult or expensive, but that the more complicated systems provide detailed information which cannot be obtained in any other way. The final chapter, Chapter 5, deals with the applications of gait analysis. This is the area in which the most progress has been made in the past few years, and in which the most progress is to be anticipated in the future. The literature of the field is heavily biased towards research rather than clinical application, but the value of the methodology is gradually coming to be realized in a number of clinical conditions.

I have deliberately avoided giving references to theses and conference proceedings, since these may be difficult to find. Chapter 1 contains no references at all, as everything in it should be easy to find in standard textbooks. I have restricted the number of references quoted in the remainder of the book, not through ignorance or laziness, but rather in an attempt to identify only the most important references on particular topics. These will in turn lead on to other references, should the reader wish to study that topic in greater depth. Those not familiar with it should ask their librarian about the Science Citation Index, which uses key references from the past to identify more recent publications in the same field.

I have used Système International (SI) units throughout this book. I make no apology for this – everyone working in this field should be using the measurement units of science, rather than those of the grocery store! However, conversions are given in Appendix 2.

Since the origins of this book are international, it is hoped that it will appeal to an international readership. It was written during my last few months at the University of Oxford, England, and my first few months at the University of Tennessee at Chattanooga, in the United States. It draws on reference material from both sides of the Atlantic, and parts of it were written on journeys across that ocean!

Michael W. Whittle
Chattanooga, Tennessee
May 1990

1

Basic Sciences

All voluntary movement, including walking, results from a complicated process involving the brain, spinal cord, peripheral nerves, muscles, bones and joints. Before considering in detail the process of walking, what can go wrong with it, and how it can be studied, it is necessary to have a basic understanding of three scientific disciplines – anatomy, physiology and biomechanics. It is hoped that this chapter will provide instruction in the rudiments of these subjects, for those not already familiar with them, and will also prove to be a convenient source of reference material.

Anatomy

It is not the intention of this book to attempt to teach in detail the anatomy of the locomotor system. This is a complex subject which is better treated in standard anatomy textbooks. The notes which follow give only an outline of the subject, but one which should be sufficient for an understanding of gait analysis. The anatomical names for the different parts of the body vary somewhat from one textbook to another – as far as possible the most common name has been used in this text. The section starts by describing some basic anatomical terms, and then goes on to describe the bones, joints, muscles, nervous system and blood supply.

Basic anatomical terms

The anatomical terms describing the relationships between different parts of the body are based on the *anatomical position*, in which a person is standing erect, with the feet together and the arms by the sides of the body, with the palms forward. This position, together with the reference planes and the terms describing relationships between different parts of the body, is illustrated in Fig. 1.1.

Six terms are used to describe directions, with relation to the center of the body. These are best defined by example:

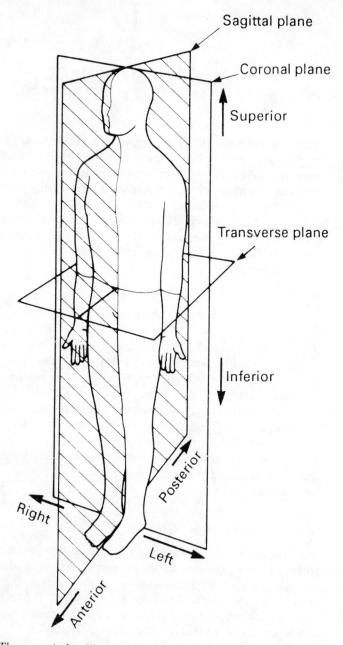

Fig. 1.1   *The anatomical position, with three reference planes and six fundamental directions.*

1. the umbilicus is *anterior*
2. the buttocks are *posterior*
3. the head is *superior*
4. the feet are *inferior*
5. *left* is self-evident
6. so is *right*.

The anterior surface of the body is *ventral* and the posterior surface is *dorsal*. The terms *cephalad* (towards the head) and *caudad* (towards the 'tail') are sometimes used in place of superior and inferior.

Within a single part of the body, six additional terms are used to describe relationships:

1. *Medial* means towards the midline of the body: the big toe is on the medial side of the foot.
2. *Lateral* means away from the midline of the body: the little toe is on the lateral side of the foot.
3. *Proximal* means towards the rest of the body: the shoulder is the proximal part of the arm.
4. *Distal* means away from the rest of the body: the fingers are the distal part of the hand.
5. *Superficial* structures are close to the surface.
6. *Deep* structures are far from the surface.

The motion of the limbs is described using reference planes:

1. A *sagittal* plane is any plane which divides part of the body into right and left portions. The *median* plane is the midline sagittal plane, which divides the whole body into right and left halves.
2. A *coronal* plane divides a body part into front and back portions.
3. A *transverse* plane divides a body part into upper and lower portions.

The term *frontal plane* is sometimes used in place of coronal plane, and the transverse plane may also be called the *horizontal plane*, although it is only horizontal when in the standing position.

There are three mutually perpendicular axes about which movement of a joint could take place, although most joints can only make some of these movements. The directions of these motions for the major joints of the legs are shown in Fig. 1.2. The possible movements are:

1. *Flexion* and *extension* – in the ankle these are called *plantarflexion* and *dorsiflexion*, respectively.

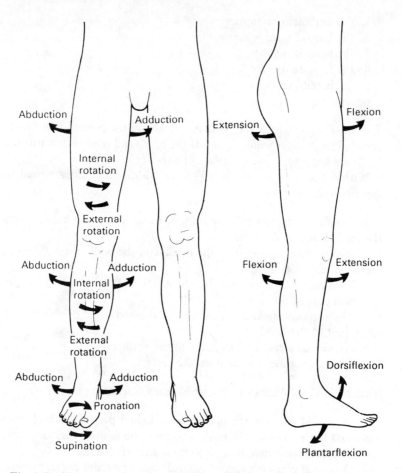

Fig. 1.2   *Movements about the major joints of the lower limb.*

2. *Abduction* and *adduction.*

3. *Internal* and *external rotation*, also called *medial* and *lateral rotation* respectively.

Other terms which are used to describe the motions of the joints or of body segments are:

1. *Varus* and *valgus*, which describe an angulation of a joint towards or away from the midline respectively; knock knees are in valgus, bow legs are in varus.

2. *Pronation* and *supination*, which are rotations about the long axis

of the hand or foot; pronation of both hands brings the thumbs together; supination of both feet brings the soles together; for the feet, the letters Ph.D (Pronation – hallux Down) are an aide-mémoire.

3. *Inversion* of the foot combines plantarflexion, supination and adduction: it takes the little toe medially and downwards. *Eversion* is the converse: it takes the little toe laterally and upwards.

It should be noted that these definitions are not unanimously agreed for the foot. Some authorities regard inversion and eversion as the basic movements, and pronation and supination as combined movements.

Bones

Although it could be argued that almost every bone in the body takes part in walking, from a practical point of view it is only necessary to consider the bones of the pelvis and legs. These are shown in Fig. 1.3.

The *pelvis* is formed from the sacrum, the coccyx and the two innominate bones. The *sacrum* consists of the five sacral vertebrae, fused together. The *coccyx* is the vestigial 'tail', made of three to five rudimentary vertebrae. The *innominate bone* on each side is formed by the fusion of three bones – the *ilium, ischium* and *pubis*. The only real movement between the bones of the pelvis occurs at the sacroiliac joint, and this movement is very small. It is thus reasonable, for the purposes of gait analysis, to regard the pelvis as being a single rigid structure. The superior surface of the sacrum articulates with the fifth lumbar vertebra of the spine. On each side of the lower part of the pelvis is the acetabulum of the hip joint, into which fits the head of the femur.

The *femur* is the longest bone in the body. The spherical femoral head articulates with the pelvic acetabulum to form the hip joint. The neck of the femur runs downwards and laterally from the femoral head to meet the shaft of the bone, which continues downwards to the knee joint. At the junction of the neck and the shaft are two bony protuberances, where a number of muscles are inserted – the greater trochanter laterally, which can be felt beneath the skin, and the lesser trochanter medially. The bone widens at its lower end to form the medial and lateral condyles. These form the upper half of the knee joint, and have a groove between them anteriorly, which articulates with the patella.

The *patella* or kneecap is a sesamoid bone. That is to say, it is embedded within a tendon – in this case the massive quadriceps tendon, the continuation of which, beyond the patella, is known as the patellar tendon. The anterior surface of the patella is subcutaneous. Its

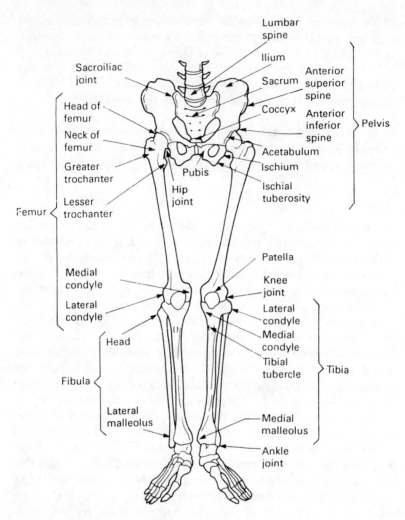

Fig. 1.3 *Bones and joints of the lower limbs.*

posterior surface articulates with the anterior surface of the lower end of the femur to form the patellofemoral joint. It has an important mechanical function, which is to displace the quadriceps tendon forwards, and thereby to improve its leverage.

The *tibia* extends from the knee joint to the ankle joint. Its upper end is broadened into medial and lateral condyles, with an almost flat upper surface which articulates with the femur. The tibial tubercle is a small bony prominence on the front of the tibia, where the patellar tendon is inserted. The anterior surface of the tibia is subcutaneous. The lower end of the tibia forms the upper and medial surfaces of the ankle joint, with a subcutaneous medial projection called the medial malleolus.

The *fibula* is next to the tibia on its lateral side. For most of its length it is a fairly slim bone, although the upper end is broadened a little to form the head. The lower end is also broadened, to form the lateral part of the ankle joint, with a subcutaneous lateral projection known as the lateral malleolus. The tibia and fibula are in contact with each other at both upper and lower ends, as the tibiofibular joints. Movements at these joints are very small, however, and they will not be considered further. A layer of fibrous tissue, known as the interosseous membrane, lies between the bones.

The foot is a very complicated structure (Fig. 1.4), which is best thought of as being in three parts:

1. The *hindfoot*, which consists of two bones, one on top of the other
2. The *midfoot*, which consists of five bones, packed closely together
3. The *forefoot*, which consists of the five metatarsals and the toes.

The *talus* or astragalus is the upper of the two bones in the hindfoot. Its superior surface forms the ankle joint, articulating above and medially with the tibia, and laterally with the fibula. Below, the talus articulates with the calcaneus through the subtalar joint. It articulates anteriorly with the most medial of the midfoot bones – the navicular.

. The *calcaneus* or os calcis lies below the talus, and articulates with it through the subtalar joint. Its lower surface transmits the body weight to the ground through a thick layer of fat, fibrous tissue and skin. The anterior surface articulates with the most inferior and lateral of the midfoot bones – the cuboid.

The midfoot consists of five bones:

1. The *navicular*, which is medial and superior
2. The *cuboid*, which is lateral and inferior
3. Three *cuneiform* bones, which lie in a row, distal to the navicular.

The five *metatarsals* lie roughly parallel to each other, the lateral two articulating with the cuboid, and the medial three with the three cuneiform bones.

The *phalanges* are the bones of the toes; there are two in the big toe and three in each of the other toes. The big toe is also called the great toe or *hallux*.

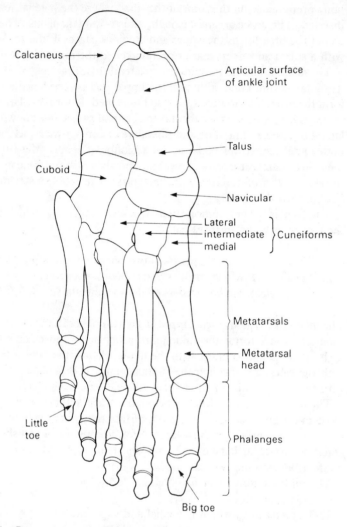

Fig. 1.4 *Bones of the right foot, from above.*

Joints and ligaments

A joint occurs where one bone is in contact with another. From a practical point of view, they can be divided into synovial joints, in which significant movement can take place, and the various other types of joint in which only small movements can occur. Since gait analysis is generally concerned only with gross movements, the description which follows deals only with synovial joints. In a *synovial joint*, the bone ends are covered in *cartilage* and the joint is surrounded by a *synovial capsule*, which secretes the lubricant *synovial fluid*. Most joints are stabilized by *ligaments*, which are bands of relatively inelastic fibrous tissue connecting one bone to another. *Fascia* is a special type of ligament, being a continuous sheet of fibrous tissue.

The *hip* joint is the only true ball-and-socket joint in the body, the ball being the head of the femur and the socket the acetabulum of the pelvis. Extremes of movement are prevented by a number of ligaments running between the pelvis and the femur, by a capsule surrounding the joint, and by a small ligament – the ligamentum teres – which joins the center of the head of the femur to the center of the acetabulum. The joint is capable of flexion, extension, abduction, adduction, internal and external rotation (see Fig. 1.2).

The *knee* joint lies between the medial and lateral condyles of the femur, above, and the corresponding condyles of the tibia, below. The articular surfaces on the medial and lateral sides are separate, making, in effect, two joints, side by side. The femoral condyles are curved both from front to back and from side to side, whereas the tibial condyles are almost flat. The 'gap' this would leave around the point of contact is filled, on each side, by a 'meniscus', commonly called a 'cartilage', which acts to spread the load and reduce the contact pressure.

The motion of the joint is controlled by five ligaments which, between them, exert very close control over the movements of the knee:

1. The medial collateral ligament (MCL), which prevents the medial side of the joint from opening up, i.e. it opposes abduction.
2. The lateral collateral ligament (LCL) which similarly opposes adduction.
3. The posterior joint capsule, which prevents hyperextension (excessive extension) of the joint.
4. The anterior cruciate ligament (ACL), in the center of the joint between the condyles, which is attached to the tibia anteriorly and the femur posteriorly, and prevents the tibia from moving forwards relative to the femur.
5. The posterior cruciate ligament (PCL), also in the center of the joint, which is attached to the tibia posteriorly and the femur anteriorly, and prevents the tibia from moving backwards relative to the femur.

The anterior and posterior cruciate ligaments are named for their positions of attachment to the tibia. They appear to act together as a 'four-bar-linkage', which imposes a combination of sliding and rolling on the joint, and moves the contact point forwards as the joint extends, and backwards as it flexes. This means that the flexion axis of the joint is not fixed, but changes with the angle of flexion or extension.

In the normal individual, the motions of the knee are flexion and extension, with a small amount of internal and external rotation. As the knee comes to full extension, there is an external rotation of a few degrees – the so-called automatic rotation or 'screw-home' mechanism.

The *patellofemoral* joint lies between the posterior surface of the patella and the anterior surface of the femur. The articular surface consists of a shallow V-shaped ridge on the patella, which fits into a shallow groove between the medial and lateral condyles. The only normal movement is the patella moving up and down the groove, during extension and flexion of the knee. This causes different areas of the patella to come into contact with different parts of the joint surfaces of the femur.

The *ankle* joint has three surfaces – upper, medial and lateral. The upper surface is the main articulation of the joint; it is cylindrical, and formed by the tibia above and the talus below. The medial joint surface is between the talus and the inner aspect of the medial malleolus of the tibia. Correspondingly, the lateral joint surface is between the talus and the inner surface of the lateral malleolus of the fibula.

The major ligaments of the ankle joint are those between the tibia and the fibula, preventing these two bones from moving apart, and the collateral ligaments on both sides – between the two malleoli and both the talus and calcaneus – which keep the joint surfaces in contact. The ankle joint, being cylindrical, has only one significant type of motion – dorsiflexion and plantarflexion.

The *subtalar* or talocalcaneal joint lies between the talus above and the calcaneus below. It has two articular surfaces, one in front of and medial to the other. Large numbers of ligaments join the two bones to each other and to all the adjacent bones. The axis of the joint is oblique, and it is difficult to describe the motion, since it takes place in more than one anatomical plane. However, from a functional point of view, the importance of the subtalar joint is that it permits abduction and adduction of the hindfoot. When performing gait analysis it is usually impossible to distinguish between movement at the ankle joint and that taking place at the subtalar joint, and it is reasonable to refer to motion at the 'ankle/subtalar complex'. This motion in normal individuals includes dorsiflexion, plantarflexion, abduction and adduction, but not rotation about the long axis of the leg.

The *midtarsal* joints lie between each of the tarsal bones and its immediate neighbors, making a very complicated structure. The movement of most of these joints is very small, as there are ligaments crossing the joints, and the joint surfaces are not shaped for large movements. As a result, the midtarsal joints may be considered together to provide a flexible linkage between the hindfoot and the forefoot, which is capable of a small amount of movement in all directions.

The *tarsometatarsal* joints, between the cuboid and the cuneiforms proximally and the five metatarsals distally, are capable of only small gliding movements, because of the relatively flat joint surfaces and the ligaments binding the tarsal bones to the metatarsals, and binding the metatarsals together. There are also joint surfaces between adjacent metatarsals, except for the medial one.

The *metatarsophalangeal* and *interphalangeal* joints consist of a convex proximal surface fitting into a shallow concave distal surface. The metatarsophalangeal joints permit abduction and adduction as well as flexion and extension. The interphalangeal joints are restricted by their ligaments to flexion and extension, the range of flexion being greater than that of extension.

No description of the anatomy of the foot is complete without mention of the arches. The bones of the foot are bound together by ligamentous structures, reinforced by muscle tendons, to make a flexible structure which acts like two strong curved springs, side by side. These are the longitudinal arches of the foot, and they cause the body weight to be transmitted to the ground primarily through the calcaneus and the metatarsal heads. The midfoot transmits relatively little weight directly to the ground because it is lifted up, particularly on the medial side. The posterior end of both arches is the calcaneus. The *medial arch* (Fig. 1.5) goes upwards through the talus, and then gradually down again through the navicular and cuneiforms to the medial three metatarsals, which form the distal end of the arch. The *lateral arch* (Fig. 1.6) passes forwards from the calcaneus through the cuboid

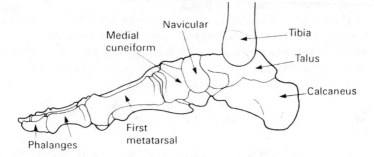

Fig. 1.5 *Medial arch of the right foot.*

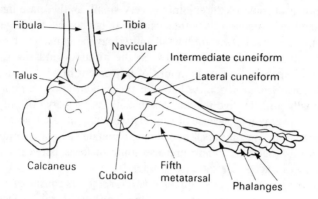

Fig. 1.6 *Lateral arch of the right foot.*

to the lateral two metatarsals. As well as the two longitudinal arches, there is also a *transverse arch*, the midline of the foot being higher than the medial and lateral borders, from the level of the navicular to the metatarsal heads.

Muscles and tendons

Muscles are responsible for movement at joints. Most muscles are attached to different bones at the two ends, and cross over either one joint (*monarticular* muscle) or two joints (*biarticular* muscle). In many cases the attachment to one of the bones covers a broad area, whereas at the other end it narrows into a *tendon*, which is attached to the other bone. It is usual to talk about a muscle as having an 'origin' and an 'insertion', although these terms are not always clearly defined. Ligaments and tendons are obviously similar and frequently confused. As a general rule ligaments make connections between bones, whereas tendons connect muscles to bones.

The account which follows gives very brief details of the muscles of the pelvis and lower limb, including their major actions. Most muscles also have secondary actions, which may vary according to the position of the joints, particularly with biarticular muscles. The larger and more superficial muscles are illustrated in Fig. 1.7.

Muscles acting only at the hip joint

1. *Psoas major* originates from the front of the lumbar vertebrae. *Iliacus* originates on the inside of the pelvis. The two tendons combine

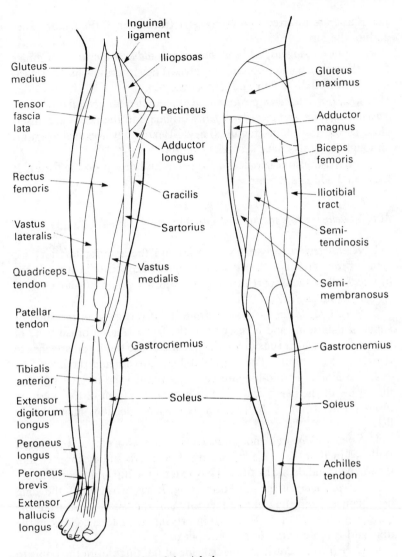

Fig. 1.7 *Superficial muscles of the right leg.*

to form the *iliopsoas*, inserted at the lesser trochanter of the femur. The main action of these two muscles is to flex the hip.

2. *Gluteus maximus* originates from the back of the pelvis and is inserted into the back of the shaft of the femur near its top. It extends the hip.

3. *Gluteus medius* and *gluteus minimus* originate from the side of the

pelvis and are inserted into the greater trochanter of the femur. They abduct the hip.

4. *Adductor magnus, adductor brevis* and *adductor longus* all originate from the ischium and pubis of the pelvis. They insert in a line down the medial side of the femur and adduct the hip.

5. *Quadratus femoris, piriformis, obturator internus, obturator externus, gemellus superior* and *gemellus inferior* originate in the pelvis and insert close to the top of the femur. They all externally rotate the femur, although most also have secondary actions.

6. *Pectineus* also originates in the pelvis and inserts on the femur. It flexes and adducts the hip.

Muscles acting across the hip and knee joints

1. *Rectus femoris* originates from around the anterior inferior iliac spine of the pelvis and inserts into the quadriceps tendon. It flexes the hip, as well as being part of the *quadriceps* – a group of four muscles which extend the knee.

2. *Tensor fascia lata* originates from the pelvis close to the anterior superior iliac spine and is inserted in the fascia lata – a broad band of fibrous tissue which runs down the outside of the thigh and attaches to the head of the fibula. The muscle abducts the hip and the knee.

3. *Sartorius* is a strap-like muscle originating at the anterior superior iliac spine of the pelvis and winding around the front of the thigh, to insert on the front of the tibia on its medial side. It is mainly a hip flexor.

4. *Semimembranosus* and *semitendinosus* are two of the *hamstrings*. Both originate at the ischial tuberosity of the pelvis and are inserted into the medial condyle of the tibia. They extend the hip and flex the knee.

5. *Biceps femoris* is the third hamstring. It has two origins – the 'long head' comes from the ischial tuberosity and the 'short head' from the middle of the shaft of the femur. It inserts into the lateral condyle of the tibia and is a hip extensor and knee flexor.

6. *Gracilis* runs down the medial side of the thigh from the pubis to the back of the tibia on its medial side. It adducts the hip and flexes the knee.

Muscles acting only at the knee joint

1. *Vastus medialis, vastus intermedius* and *vastus lateralis* are the three remaining elements of the quadriceps muscle. They all originate from

the upper part of the femur, on the medial, anterior and lateral sides respectively, and all combine (with the rectus femoris) to become the quadriceps tendon. This surrounds the patella, and continues beyond it as the patellar tendon, which inserts into the tibial tubercle.

2. *Popliteus* is a small muscle behind the knee. It flexes and helps to unlock the knee by internally rotating the tibia at the beginning of flexion.

Muscles acting across the knee and ankle joints

1. *Gastrocnemius* originates from the back of the medial and lateral condyles of the femur. Its tendon joins with that of the soleus (and sometimes also the plantaris) to form the *Achilles tendon*, which inserts into the back of the calcaneus. The main action of these muscles is to plantarflex the ankle, although the gastrocnemius is also a flexor of the knee.

2. *Plantaris* is a very slender muscle running deep to the gastrocnemius from the lateral condyle of the femur to the calcaneus. It is a feeble plantarflexor of the ankle. The soleus, gastrocnemius and plantaris together are called the *triceps surae*.

Muscles acting across the ankle and subtalar joints

1. *Soleus* arises from the posterior surface of the tibia, fibula and the deep calf muscles. Its tendon joins with that of the gastrocnemius to plantarflex the ankle.

2. *Extensor hallucis longus, extensor digitorum longus, tibialis anterior* and *peroneus tertius* form the anterior tibial group. They originate from the anterior aspect of the tibia and fibula and the interosseous membrane. All serve to dorsiflex the ankle. The former two are inserted into the toes, which they extend; the latter two are inserted into the tarsal bones and raise the midfoot on the medial side (tibialis anterior) or lateral side (peroneus tertius).

3. *Flexor hallucis longus, flexor digitorum longus, tibialis posterior, peroneus longus* and *peroneus brevis* are the deep calf muscles, and all arise from the tibia, fibula and interosseous membrane. They all plantarflex the ankle. Flexor hallucis longus and flexor digitorum longus are flexors of the toes. The peronei are on the lateral side and evert the foot; tibialis posterior is on the medial side and inverts the foot.

Muscles within the foot

1. *Extensor digitorum brevis* and the *dorsal interossei* are on the dorsum of the foot. The former muscle extends the toes and the latter muscles abduct and flex the toes.

2. *Flexor digitorum brevis, abductor hallucis* and *abductor digiti minimi* form the superficial layer of the sole of the foot. They flex the toes and abduct the big toe and the little toe, respectively.

3. *Flexor accessorius, flexor hallucis brevis* and *flexor digiti minimi brevis* form an intermediate layer in the sole of the foot. Between them they flex all the toes.

4. The *adductor hallucis* is in two parts – the oblique and transverse heads. It adducts the big toe.

5. The *plantar interossei* and the *lumbricals* lie in the deepest layer of the sole of the foot. The former adduct and flex the toes; the latter flex the proximal phalanges and extend the distal phalanges.

The above five groups of muscles are known collectively as the *intrinsic muscles* of the foot.

Spinal cord and spinal nerves

The *spinal cord* is an extension of the brain, and plays an active role in the processing of nerve signals. Like the brain itself, it consists of white matter, which is bundles of nerve fibers, and gray matter, which contains many cell bodies and nerve endings, where the synapses (connections) between nerve cells take place (Fig. 1.8). The spinal cord lies within the spinal canal, which is formed in front by the vertebral bodies and behind by the neural arches of the vertebrae. The vertebrae are divided into four groups – cervical (7), thoracic (12), lumbar (5) and sacral (5). It is usual to use abbreviated names, e.g. the fourth thoracic vertebra is known as T4.

The spinal cord is shorter than the spinal canal, terminating in adults at approximately the level of the first lumbar vertebra (L1), and a little lower in children. Beyond the end of the spinal cord is a bundle of nerves known as the *cauda equina*, which consists of those nerve roots which enter and leave the lower levels of the spinal canal (Fig. 1.9). There are eight cervical nerve roots, but only seven cervical vertebrae, each nerve root except the eighth emerging above the correspondingly numbered vertebra. In the remainder of the spine, the nerve roots emerge below the corresponding vertebrae.

The organization of the *neurons* (nerve cells) of the spinal cord and

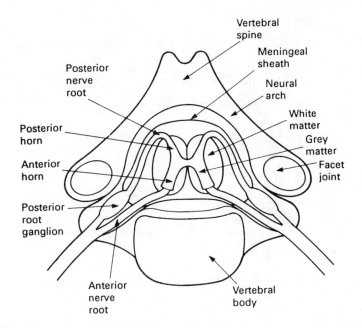

Fig. 1.8 *Cross-section of spinal cord and vertebra.*

the peripheral nerves is extremely complicated. It is possible to give here only a brief outline, although further details will be given in the physiology section later in this chapter. The main neurons responsible for muscle contraction pass down from the brain as *upper motor neurons* in the 'descending' tracts of the spinal cord. At the appropriate spinal level they enter the gray matter and connect with the *lower motor neurons*, also called *efferent* neurons. The axons (nerve fibers) of these cells pass out of the spinal cord through the *anterior root*, combine with other spinal roots and then split into smaller and smaller nerves, finally reaching the muscle itself.

Nerve fibers also pass in the opposite direction, from the muscles, skin and other structures to the spinal cord. They enter the spinal cord at the *posterior root*, having passed through the *posterior root ganglion*, a swelling which contains the cell bodies of the neurons. These *afferent* neurons transmit many different types of sensory information. Some connect with the nerve fibers which pass up the spinal cord to the brain in the 'ascending' tracts, while others synapse with other nerve cells at the same or nearby spinal levels. Connections within the spinal cord are responsible for the spinal reflexes, which will be referred to later.

When the spinal cord is damaged by accident or disease, the results

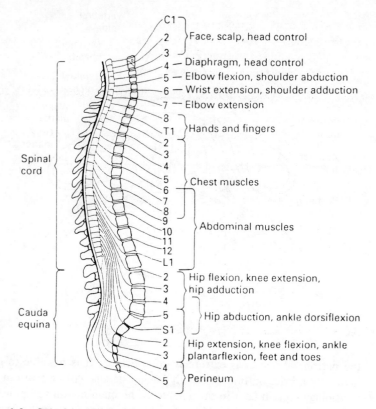

C1 ⎤
2 ⎦ Face, scalp, head control
3
4 — Diaphragm, head control
5 — Elbow flexion, shoulder abduction
6 — Wrist extension, shoulder adduction
7 — Elbow extension
8 ⎤
T1 ⎦ Hands and fingers
2
3
4
5 ⎤
6 ⎦ Chest muscles
7
8
9 ⎤
10 ⎦ Abdominal muscles
11
12
L1
2 ⎤ Hip flexion, knee extension,
3 ⎦ hip adduction
4
5 ⎤ Hip abduction, ankle dorsiflexion
S1 ⎦
2 ⎤ Hip extension, knee flexion, ankle
3 ⎦ plantarflexion, feet and toes
4
5 ⎤ Perineum

Spinal cord

Cauda equina

Fig. 1.9   *Spinal cord and spinal nerves, with main functions served.*

depend both on the spinal level at which the damage occurred, and on whether the cord was totally or partially transected (cut through). A wide variety of disabilities may result from incomplete destruction of the spinal cord. If the cord is totally transected, the upper motor neurons are unable to control the muscle groups at or below that level, so voluntary control of those muscles is lost. There is also a total loss of sensation below the level of the lesion. However, at levels below the damaged area, there is usually preservation of the lower motor neurons, the sensory nerves and the spinal reflexes. Injury to the vertebral column below L1 will damage the cauda equina, rather than the spinal cord itself. The cauda equina consists of lower motor neurons and sensory fibers, and damage to it produces a totally different clinical picture from damage to the upper motor neurons.

Patients paralyzed at the level of the cervical spine are *tetraplegic* or *quadriplegic*, with paralysis of the arms and legs. With a cervical lesion above C4 the diaphragm is also paralyzed, making breathing difficult or impossible, and the chances of survival are poor. At the lower cervical levels some arm or hand function is preserved. Where the spinal cord damage is at thoracic or lumbar level, only the legs are paralyzed and the patient is *paraplegic*. Where only the cauda equina is damaged, the patient has an incomplete paraplegia, and may be able to walk wearing callipers. Patients with paralysis restricted to one side of the body are *hemiplegic*. Sometimes the suffix '-paretic' is used in place of '-plegic' to imply a *partial* paralysis.

The area of skin served by the sensory nerves from a particular spinal root is known as a *dermatome*. The distribution of the dermatomes for all the spinal nerves is shown in Fig. 1.10. In the legs, the anterior surface is innervated by the higher spinal segments and the posterior by the lower ones; loss of sensation from the buttocks and perineum is likely to follow spinal injury at almost any level.

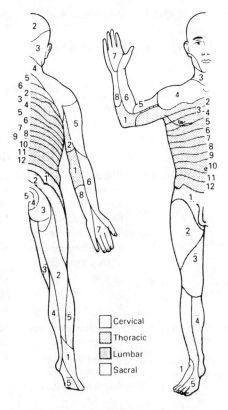

Fig. 1.10   *Sensory distribution of spinal nerve roots.*

Peripheral nerves

On emerging from the spinal cord, the spinal roots from adjacent levels form a network known as a *plexus*. The peripheral nerves which emerge from such a plexus usually contain nerve fibers from several adjacent spinal roots. The peripheral nerves supplying the muscles of walking all arise from either the *lumbar plexus* or the *sacral plexus*. Table 1.1 gives brief details of the motor and sensory distribution of the nerves arising from the lumbar plexus, and Table 1.2 the corresponding information for the sacral plexus (sometimes called the lumbosacral plexus).

Table 1.1 *Distribution of nerves arising from lumbar plexus*

Nerve	Origin	Motor	Sensory
Anterior lumbar nerves	L2–L3	Psoas major	—
Iliohypogastric	T12–L1	Abdominal wall	Abdominal wall Lateral buttocks
Ilioinguinal	T12–L1	Abdominal wall	Abdominal wall Upper thigh Genitalia
Genitofemoral	L1–L2	Genitalia	Upper thigh (anterior) Genitalia
Lateral femoral cutaneous	L2–L3	—	Upper thigh (lateral)
Femoral	L2–L4	Iliacus Pectineus Sartorius Rectus femoris Vastus lateralis Vastus intermedius Vastus medialis	Anterior thigh Medial thigh Medial leg Medial foot Hip joint Knee joint
— Saphenous	L2–L4	—	Medial leg Medial foot Knee joint
Obturator	L2–L4	Obturator externus Pectineus Adductor longus Adductor brevis Adductor magnus Gracilis	Medial thigh Hip joint Knee joint

Table 1.2 *Distribution of nerves arising from sacral plexus*

Nerve	Origin	Motor	Sensory
Superior gluteal	L4–S1	Gluteus minimus Gluteus medius Tensor fascia lata	—
Inferior gluteal	L5–S2	Gluteus maximus	—
Nerve to piriformis	S1–S2	Piriformis	—
Nerve to quadratus femoris	L4–S1	Quadratus femoris Inferior gemellus	Hip joint
Nerve to obturator internus	L5–S2	Obturator internus Superior gemellus	—
Perforating cutaneous	S2–S3	—	Medial buttock
Posterior cutaneous	S1–S3	—	Inferior buttock Posterior thigh Upper calf
Sciatic	L4–S3	Biceps femoris Semimembranosus Semitendinosus Adductor magnus	Knee joint
— Tibial	L4–S3	Gastrocnemius Plantaris Soleus Popliteus Tibialis posterior Flex. dig. longus Flex. hall. longus	Lower leg (posterior) Posterior foot Lateral foot Knee joint Ankle joint
— Medial plantar		Abductor hallucis Flex. dig. brevis Flex. hall. brevis	Medial foot Distal toes Tarsal joints
— Lateral plantar		Remaining muscles of foot	Lateral foot Tarsal joints
— Common peroneal	L4–S2	—	Knee joint
— Superficial peroneal		Peroneus longus Peroneus brevis	Anterior leg Dorsal foot
— Deep peroneal		Tibialis anterior Ext. hall. longus Ext. dig. brevis Ext. dig. longus Peroneus tertius	Great toe Second toe Ankle joint Tarsal joints
Pudendal	S2–S4	Perineum	Genitalia

Blood supply

The blood supply to the lower limb is likely to be of interest in gait analysis only if it is inadequate. Damage to an artery or occlusion by atherosclerosis or blood clot may have two consequences. If blood flow to a muscle or group of muscles is reduced for any reason, exercise will result in pain due to inadequate oxygenation. If this arises during walking, and recovers when the subject stops, it is known as *intermittent claudication*. Total loss of blood supply to one or more muscles will lead to death of some or all of the muscle tissue. This will lead to loss of the use of the muscle, and may be followed by shortening of the dead muscle as it is replaced by fibrous tissue – an *ischaemic contracture*. Damage or destruction of the blood supply may also cause destruction of the peripheral nerves, with a consequent loss of both sensory and motor functions.

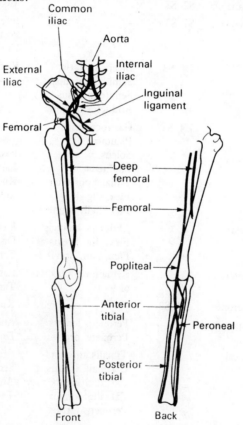

Fig. 1.11 *Major arteries of the legs.*

A simplified diagram of the blood supply of the legs is shown in Fig. 1.11. The main blood vessel of the body, the *aorta*, divides into the two *common iliac* arteries in front of L4. The common iliac divides into the *internal iliac*, which supplies organs in the pelvis, and the *external iliac*, which passes beneath the inguinal ligament to enter the leg, at which point its name changes to the *femoral* artery. The *deep femoral* artery is given off high in the thigh. The femoral artery passes down the thigh and behind the femur, changing its name to the *popliteal* artery. After passing behind the knee, the *anterior tibial* artery is given off, which passes between the tibia and fibula into the anterior compartment. The popliteal artery divides to become the *posterior tibial* on the medial side, and the *peroneal* on the lateral side. Around the ankle and in the foot there is considerable interconnection between the terminal branches of the anterior and posterior tibial and the peroneal arteries.

The veins of the leg are more variable than the arteries. There are usually two superficial veins – the *long saphenous* on the front of the leg and the *short saphenous* on the back of the calf – and a number of deep veins including the *popliteal*, which becomes the *femoral* as it passes up the leg, and pairs of veins known as *venae comitantes*, which accompany the arteries.

Physiology

Nerves

Mention has already been made of the nerve cell or *neuron*, the basic element of the nervous system. Although there is considerable variation in the structure of neurons in different parts of the nervous system, they all consist of four basic elements (Fig. 1.12) – *dendrites*, *cell body*, *axon* and *presynaptic endings*. Nerve impulses are conducted from the dendrites to the cell body, and thence down the axon to the presynaptic endings. These contain very small packages of a chemical known as a *neurotransmitter*, which is released and crosses a small space known as a *synapse*, to stimulate the dendrite of another neuron. Within the brain and spinal cord, dendrites are stimulated to produce a nerve impulse by the axons of other cells, and in turn the nerve impulse sent along the axon stimulates the dendrites of other neurons. The peripheral nerves contain *motor neurons*, whose axons stimulate muscle fibers, and *sensory neurons*, in which the dendrites are stimulated by the sense organs.

The brain and spinal cord consist of millions of neurons, connected together in a vast and complex network. A single peripheral nerve

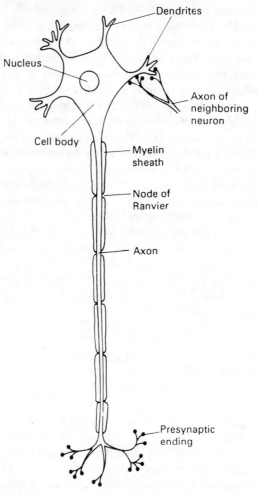

Fig. 1.12 *Neuron structure.*

consists of the axons and dendrites of hundreds or even thousands of individual neurons. The tissues of the brain, spinal cord and peripheral nerves also contain a number of other types of cell known as *neuroglia*, whose functions are either to provide physical support for the neurons or to perform a variety of maintenance functions.

The *upper motor neurons* arise in several different areas of the brain, but most notably in the motor cortex, and pass down the spinal cord to the appropriate level, crossing over to the other side at some point in the journey. Within the anterior horn of gray matter, the upper motor neurons synapse with the lower motor neurons, as well as with a large number of other neurons, that take part in the complex system of motor control.

The *lower motor neurons* or efferent neurons arise in the anterior horns of the spinal cord, emerge from the anterior spinal roots and pass down the peripheral nerves to the muscles. The axon usually branches at its distal end, where it synapses with the muscle cells. The nerve impulse causes contraction of the muscle, by a process which will be explained later.

The *sensory nerves* arise in the sense organs of the skin, joints, muscles and other structures. The sense organ itself stimulates the end of the dendrite of the afferent neuron. The dendrite usually commences as a number of branches; these come together and run up the peripheral nerve to enter the posterior root of the spinal cord. The cell body is in the *posterior root ganglion* (see Fig. 1.8), and the axon runs from the ganglion into the spinal cord itself, usually terminating in the posterior horn of gray matter, where it synapses with other neurons. As well as the familiar sensations of touch, temperature, pain and vibration, sensory nerves also carry *proprioception* signals, which are used for feedback in the control of the limbs. These signals include the positions of the joints and the tension in the muscles and ligaments.

The term *nerve impulse* has been used several times without explanation, and it is now time to rectify this deficiency. The nature of the nerve impulse is a little difficult to grasp, since it is a complicated electrochemical process.

There are different concentrations of ions between the inside of all types of cell and the surrounding extracellular fluid. (Ions are atoms or molecules that have gained or lost electrons, making them electrically charged.) The outside layer of a cell is known as the *cell membrane*; it is largely impermeable to sodium ions, and any that leak in are 'pumped' out again. The inside of the cell contains large negatively charged ions, such as proteins, which are unable to pass through the cell membrane. The high concentration of sodium ions outside the cell, and of negative ions inside it, causes an automatic compensation, which results in a high concentration of potassium ions on the inside of the cell, and of chloride ions outside. The inside of the cell thus has higher concentrations of potassium and large negative ions, while the outside has more sodium and chloride. The result of these imbalances in ionic concentration is a voltage difference between the inside and outside of the cell, across the thickness of the cell membrane. This *membrane potential* can be measured if a suitably small electrode is inserted. The normal resting membrane potential for a neuron is around $-70\,\text{mV}$, the negative sign indicating that the inside of the cell is negative with respect to the outside.

All body cells exhibit a membrane potential, but nerve and muscle cells differ from other cells in that they can manipulate it by altering the permeability of the cell membrane to sodium and potassium ions. This is the mechanism by which both nerve impulses and muscular contraction are propagated. If the membrane potential is lowered by about 20 mV, the membrane suddenly becomes extremely permeable to sodium ions, which enter rapidly from the extracellular fluid. While these ions are entering, the membrane potential is reversed to about +40 mV and it is said to be *depolarized*. The increase in permeability to sodium ions is short lived, and is followed by an increased permeability to potassium ions, which leave the cell, thus restoring the ionic balance and returning the membrane potential to −70 mV. The actual numbers of ions crossing the cell membrane are small, and the overall composition of the cell is not affected to any appreciable extent. It is only when the sodium ions have entered, but before the potassium ions have left, that the membrane potential is reversed. The change in membrane potential by around 110 mV, from −70 mV to +40 mV, is known as an *action potential*.

Under normal circumstances, an action potential in a neuron begins in the synapses, in response to the neurotransmitters released from the presynaptic endings of the axons of other neurons. Some of these are excitatory, which means that they reduce the membrane potential, and some are inhibitory, in that they increase it. This combination of excitatory and inhibitory influences permits the addition and subtraction of nerve impulses. If the net effect of the various excitatory and inhibitory influences causes the membrane potential to fall to around −50 mV, an action potential will occur in that region of the neuron. This action potential spreads from its origin, crossing the cell body and running down the axon to its termination.

The action potential is an 'all-or-none' phenomenon, its size and shape being independent of the intensity of the stimulus, so long as it is above the threshold – there are no 'larger' or 'smaller' action potentials. However, a nerve can pass action potentials either one after another in quick succession, or only occasionally and separated by long pauses. Thus it is the frequency of the nerve impulses, not their size, that carries the information on how hard the muscle is to be contracted, for example, or on the temperature of the skin.

Figure 1.13 shows an action potential passing along an axon from left to right. At its leading edge, sodium ions enter the axon, producing a region with reversed polarity. At the trailing edge of the action

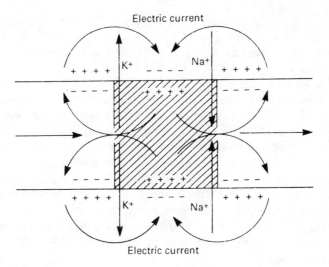

Fig. 1.13   *Propagation of action potential along axon.*

potential, potassium ions leave the axon and the membrane potential is restored. The depolarized region has a membrane potential of +40 mV, whereas the surrounding regions have a membrane potential of around −70 mV. This is equivalent to a battery producing 110 mV, and an electric current flows between the depolarized region and the surrounding normal regions of cell membrane. The passage of this electric current causes a drop in the membrane potential sufficient to generate an action potential in the normally polarized region in front, enabling the action potential to spread along the nerve. The region immediately behind an action potential becomes *refractory*, meaning that it cannot be stimulated again for a few milliseconds, so the action potential only moves in one direction.

The description so far has ignored the fact that many nerve fibers, particularly those required to send impulses quickly over long distances, are enclosed in a *myelin sheath*, as shown in Fig. 1.12. Myelin is a fatty substance which surrounds nerve fibers, both axons and dendrites, as a series of sleeves, with gaps between them known as *nodes of Ranvier*. Since myelin is an insulator, it prevents the electric current from passing through the cell membrane close to an area of depolarization, forcing it instead to pass through the next node of Ranvier, some distance along the nerve fiber. The effect of this is to cause the action potential to pass down the fiber in a series of jumps, known as *saltatory conduction*, which is much faster than the continuous propagation seen in unmyelinated fibers. A number of neurological diseases, most notably multiple sclerosis, are associated with loss of myelin from nerve fibers, with serious consequences to the functioning of the nervous system.

The speed at which nerve impulses travel depends on two things – the diameter of the nerve fiber and whether or not it is myelinated. Three speeds of fiber are found within the nervous system, known as types A, B and C. Type A fibers are all myelinated, and are subdivided by their conduction velocities into three: alpha (α) – about 100 m/s; beta (β) – about 60 m/s; and gamma (γ) – about 40 m/s. Type B and C fibers are unmyelinated with conduction velocities around 10 m/s and 2 m/s respectively. The type A fibers are the most important in gait analysis, especially the alpha fibers, which are used for the motor nerves to muscles and the faster sensory nerves such as touch. The gamma fiber is of particular importance in muscle physiology and will be referred to again later.

When an unmyelinated nerve fiber becomes damaged, recovery of function is usually impossible, because of the formation of scar tissue. For this reason, very little recovery of neuronal function takes place following damage to the brain or spinal cord, although function may be partially restored by the use of alternative neurological pathways. Myelinated fibers can recover, providing the cell body remains alive and the myelin sheaths remain in line: the nerve fiber regrows down the sheath at the rate of a few millimeters per week. In practice, if a complete nerve is divided and reconnected, most of the nerve fibers will enter the wrong myelin sheaths, although a sufficient number may be correctly connected to lead to useful sensory and motor function.

Muscles

The human body contains three types of muscle – smooth, cardiac and skeletal. The description which follows is of *skeletal muscle*, which is responsible for the movement of the limbs. It is also known as voluntary or striated muscle.

The basic unit of muscle tissue is the *muscle fiber*, each of which is a large multinucleated cell. The muscle fibers are grouped together in bundles known as *fascicles* (Fig. 1.14). The fiber is itself made up of hundreds of *myofibrils*, which have a characteristic striped appearance. These striations are due to a regular arrangement of *filaments*, which are made of two types of protein – *actin* and *myosin*. It is the sliding of these filaments past each other, by the formation and destruction of cross-bridges, which is responsible for muscle contraction.

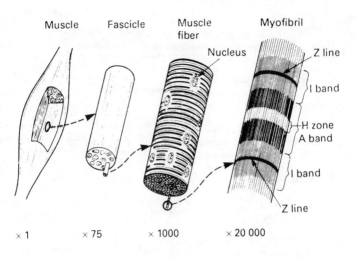

Fig. 1.14 *Macro- and microscopic structure of muscle.*

The various light and dark bands in the myofibril are identified by letters. The thin, dark 'Z' line is the origin of the slender actin filaments (Fig. 1.15). These are interleaved with the thicker myosin filaments, which form the 'A' band. The 'I' band and 'H' zone change their width during muscular contraction, as they represent the areas where the actin and myosin, respectively, are not overlapped – they were named before the process of contraction was understood!

There is an extremely complicated arrangement of membranes surrounding the myofibrils within the muscle fiber. It is responsible for the transport of nutrients and waste products, and the transmission of the muscle action potential. Outside the muscle fibers are the blood capillaries and the terminal branches of the motor nerves, each of which terminates in a *motor endplate*, which makes a synapse-like connection to a muscle fiber at a neuromuscular junction. On average, a single motor nerve will connect to about 150 muscle fibers, the combination of the neuron and the muscle fibers it innervates being known as a *motor unit*.

When an action potential passes down a nerve to the endplate, it results in the release of the transmitter substance *acetylcholine (ACh)*. This depolarizes the cell membrane of the muscle fiber, and causes a spreading wave of depolarization. As this *muscle action potential* spreads throughout the muscle fiber, it causes the release of calcium ions, which are the trigger for muscle contraction. Cross-bridges form between the actin and myosin filaments, pulling them together. The tension is

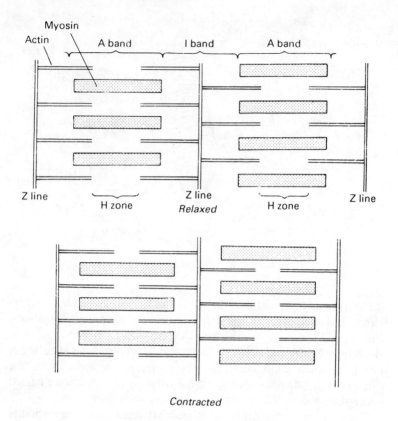

Fig. 1.15 *Sliding of filaments during muscular contraction.*

maintained for a brief period, then released if no further action potential occurs. The electrical activity of muscle action potentials can be detected, and is known as the *electromyogram (EMG)*.

The energy for muscular contraction comes from the release of a high-energy phosphate group by a chemical known as *adenosine triphosphate (ATP)*. The regeneration of ATP requires the expenditure of metabolic energy, and a failure to keep up with the demand results in muscle fatigue. There are two metabolic pathways involved in regenerating ATP. One uses up chemicals stored within the cell, without the need for oxygen, and is known as *anaerobic*; the other requires oxygen and nutrients to enter the muscle fiber from the bloodstream, and is known as *aerobic*. Anaerobic processes are quickly exhausted, although they can provide brief bursts of powerful contraction. For more sustained muscular effort, aerobic metabolism is required. Following anaerobic respiration, a muscle will have an *oxygen debt*, which will need to be 'repaid' by aerobic respiration, to remove lactic acid which accumulates in the muscle.

If a single action potential stimulates the muscle, it will respond, after a short pause known as the *latent period*, with a brief contraction known as a *twitch* (Fig. 1.16). A second action potential occurring during the latent period has no effect. If the nerve is repeatedly stimulated, but there is sufficient time for one twitch to end before the next occurs, the force of contraction will increase over the first few twitches, a phenomenon known as *treppe*. If there is insufficient time for the muscle to relax before it is stimulated again, the force will build up as a *tetanus*. (The disease of the same name results in muscular contraction due to bacterial toxins.) The muscular contractions seen in gait are all tetanic. The force which a muscle is able to generate in a tetanic contraction depends on a number of factors, particularly the strength of stimulation and the length of the muscle. The greatest force is usually produced when the muscle length is somewhere in the middle of its working range; the force of contraction falls to zero if the muscle shortens to its minimum length.

There are three different types of muscle fiber, known as I, IIa and IIb. The type of muscle fiber depends on the type of stimulation reaching it down the motor nerve, and all the fibers in a single motor unit are of the same type. *Type I fibers* are dark in color, they contract and relax slowly, and are fatigue resistant. They are used primarily for the sustained contraction used for posture control. *Type IIa fibers* are pale in color, fast to contract and relax, and easily fatigued. They are mainly used for brief bursts of powerful contraction. *Type IIb fibers* are similar to IIa, but even faster and more easily fatigued. They are only used in very strong muscular contractions. Depending on their function in the body, different muscles have different proportions of fast and

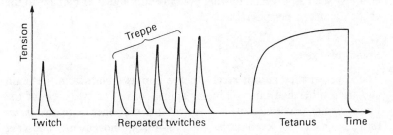

Fig. 1.16 *Response of a single muscle fiber to single stimulation, and to repeated stimulation at low and high frequencies.*

slow fibers; this is also seen in the differences between 'red meat' and 'white meat' in poultry. A change in the stimulation pattern will cause a change in the fiber type in the course of a few weeks, despite the fact that there are differences between the fiber types in the actual structure of the myosin. This ability to alter the fiber type becomes very important when electrical stimulation is used on paralyzed muscles.

When a muscle contracts, not all the motor units are active at the same time. If a stronger contraction is needed, further motor units are brought into use, a process known as *recruitment*.

If a muscle generates tension without changing its length, the contraction is known as *isometric*. If the muscle changes its length but the force of contraction remains the same, the contraction is *isotonic*. One normally thinks of a muscle shortening as it contracts – a *concentric contraction*. However, it is quite usual, particularly in gait, for a muscle to produce tension while it is lengthening – an *eccentric contraction*. For example, the quadriceps undergoes an eccentric contraction as you sit down. The muscle which is responsible for a particular action is known as an *agonist*. If two or more muscles act together, they are known as *synergists*. Muscles which oppose agonists are known as *antagonists*. As a general rule, contraction of one set of muscles results in *reciprocal inhibition* of opposing muscles.

Muscle atrophy is the term used to describe the loss of bulk and strength of a muscle when it is not used. If the motor nerve is intact, the muscle fibers will become smaller but their numbers remain the same, and subsequent restoration of muscle stimulation will lead to a full recovery of function. This happens, for example, when a limb is encased in plaster following a fracture. This type of muscle atrophy also occurs in spinal cord transection where the upper motor neurons are destroyed but the lower motor neurons remain intact. In contrast, if the lower motor neuron is destroyed, the muscle fibers shrink and become replaced by fibrous tissue. This leads to an irreversible form of muscle atrophy, such as is seen following poliomyelitis and after damage to the cauda equina or peripheral nerves.

Spinal reflexes

The lower motor neurons receive nerve impulses both from the brain and from other neurons within the spinal cord. The two areas of the brain chiefly concerned with posture and movement are the *motor cortex*, which is responsible for voluntary movement, and the *cerebellum*, which is responsible for generating patterns of muscular activity. Within the spinal cord itself, the influences of other neurons give rise to the *spinal reflexes*. There are also pattern generators for each limb within the spinal cord, which are capable of producing alternating flexion and extension.

The brain and higher centers exert an inhibitory influence on spinal reflexes, which are often very weak in normal individuals. However, the reflexes may become very strong in patients who have suffered from damage to the brain or spinal cord, due to the loss of this inhibition.

One of the most important spinal reflexes is the *stretch reflex*, which is responsible for the knee-jerk when the patellar tendon is struck by a small hammer. When a muscle is stretched, *stretch receptors* within it are stimulated, sending nerve impulses to the spinal cord along fast sensory neurons. Within the cord, these neurons synapse with and stimulate the lower motor neurons of the same muscle, causing it to contract. The stretch receptors are within the *muscle spindles*, and attached to the very small *intrafusal muscle fibers*, which are innervated by thin, relatively slow *gamma motor neurons*. These adjust the length of the spindle as the main muscle contracts and relaxes, so that it continues to work over the complete range of muscle lengths. The intrafusal fibers are also able to alter the 'sensitivity' of the stretch receptor. The stretch reflex provides a feedback system for maintaining the position of a muscle despite changes in the force applied to it.

The stretch reflex is unusual in that it involves only a single synapse, between the sensory and motor neurons, making it a *monosynaptic reflex*. Most reflexes are *polysynaptic*, involving many intermediate neurons, and often involving neurons on both sides of the spinal cord and at more than one spinal level.

Partly because of the stretch reflex, and partly through a continuous low level of activity in the motor neurons, most muscles show a certain amount of resistance to being stretched – this is known as *muscle tone*. In some individuals this effect is exaggerated, giving the clinical condition of *spasticity*, in which muscle tone is very high, small movements of the limb being opposed by strong muscular contractions. Spasticity is an important cause of gait abnormalities. It usually results from the loss of some or all of the inhibitory influence of the higher centers on the spinal reflexes, and is particularly seen in cerebral palsy or below an area of spinal cord damage.

Many different types of sensory organ in the tissues are responsible for spinal reflexes. Those of particular importance in gait analysis are the muscle spindle and the *Golgi organ*. The latter is a stretch receptor in tendons, which inhibits muscular contraction if the force applied to the tendon, either actively or passively, becomes dangerously large. Pain receptors in the limb may elicit the *flexor withdrawal reflex*, in which the flexor muscles contract and the extensors relax, hopefully to remove the limb from whatever is causing the pain. There is also a *crossed extensor reflex*, where contraction of flexors on one side is accompanied by contraction of extensors on the other. However, this is very weak in humans, even after spinal cord transection.

Biomechanics

Biomechanics is a scientific discipline which studies biological systems, such as the human body, by the methods of mechanical engineering. Since gait is a mechanical process which is performed by a biological system, the methods of biomechanics are ideally suited to its investigation. Mechanical engineering is a vast subject, but the descriptions which follow are limited to those aspects which are most relevant to gait analysis, especially mass, force, center of gravity, moments of force, and motion, both linear and angular. The science of biomechanics can be extremely mathematical, but the basic principles are easy to grasp, and the section ends with a worked example to illustrate this.

Mass

As we all live in the earth's gravitational field, we normally use the terms mass and weight to mean much the same thing. However, there is a clear distinction between them. The *mass* of an object is the amount of matter contained in it, which does not depend on whether any gravity is present, whereas *weight* is the force exerted by gravity on the object. For example, in an orbiting spacecraft there is no gravity, and all objects are weightless, although they still have mass. This means that you are still likely to be injured if someone throws a rock at you, even though it doesn't 'weigh' anything! We casually talk about measuring our body 'weight' in pounds or kilograms, but this is incorrect in scientific terms, since these are units of mass, not of force.

Force

We are all familiar in general terms with the concept of force, but the scientist uses the term in a particular way. Force is a *vector* quantity, which means that it has both magnitude and direction, in contrast to *scalar* quantities, such as temperature, which have only magnitude. In scientific texts, it is usual to print vectors in bold typeface. The Système

International (SI) unit of force is the *newton* (N). The force applied by normal earth gravity to a mass of 1 kg is 9.81 N; one newton is the force exerted by a mass of about 102 g. The earlier imperial and metric units of force were confusing and are best avoided – conversions will be found in Appendix 2. The direction of a force vector may be stated in any convenient manner, for example 20 N downwards, or 140 N at 30 degrees to the *x*-axis. However, the direction should never be omitted, unless it is obvious.

The whole science of mechanical engineering is based on the three laws of force propounded by Sir Isaac Newton, which may be paraphrased as follows:

Newton's First Law: A body will continue in a state of rest or of uniform motion in a straight line unless it is acted upon by an external force.

Newton's Second Law: An external force will cause a body to accelerate in the direction of the force. The acceleration (a) is equal to the size of the force (F) divided by the mass (m) of the object, as in the equation:

$$a = F/m$$

Newton's Third Law: To every action there is a reaction, which is equal in magnitude and opposite in direction.

Neglecting the strange behavior of atomic and subatomic particles, all physical systems must obey all three of Newton's laws, at all times.

It is easy to remember which law is which if you first imagine a brick, just floating in space (First Law); then someone pushes it, and it accelerates (Second Law); then someone stops it accelerating by pushing from the other side (Third Law). This image is not strictly accurate in mechanical terms, but it is good enough as an aid to memory!

It is easy to see that a single force acting in one direction can be balanced out by an equal force acting in the opposite direction. A much more common situation, however, is to have a number of forces acting in different directions, which together balance each other out. Providing direction is taken into account, it is possible to add and subtract force vectors – as it is with any other vectors such as velocity or acceleration. To understand how this is possible, it is necessary to appreciate the fact that a single force, acting in a single direction, can be exactly equivalent to a number of different forces acting in other directions. Conversely, any number of separate forces can be represented by an appropriate single force.

The technique used to convert a single force into two forces, acting in

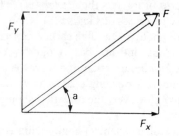

Fig. 1.17   *Resolution of force F into two components:* F_x *and* F_y.

different directions, is known as *resolving into components*. Figure 1.17 shows how the force F can be represented by two smaller forces, F_x and F_y, acting at right angles to each other. The magnitude of these forces are given by the formulae:

$$F_x = F \times \cos a$$
$$F_y = F \times \sin a$$

where a is the angle between F and F_x.

The converse process, of combining F_x and F_y to produce F, can be performed using these formulae:

$$\text{Magnitude: } F = \sqrt{(F_x{}^2 + F_y{}^2)}$$
$$\text{Angle: } a = \tan^{-1}(F_y/F_x)$$

In order to combine forces, they are first resolved into components, using a common system of directions. All the x components are added together, and so are all the y components. The resulting totals are then used to find the single equivalent force. This is illustrated in Fig. 1.18, where two forces, A and B, are combined to give a resultant force R. First, A and B are resolved into components (A_x, A_y, B_x and B_y). The algebraic sum of the x components gives the x component of the resultant, and similarly with the y components. Since A_x and B_x are in opposite directions, their algebraic sum is actually their difference. The resultant, R, is then obtained by recombining the x and y components.

When resolving a force into components, it is not necessary that the two components are at right angles, although it makes the calculations much easier, since the right-angled triangle lends itself to simple geometrical and trigonometrical methods.

Figure 1.19 shows how graphical methods can be used to combine two forces, A and B, to give a single resultant force R. The two forces are drawn as a scale diagram, with the correct angle between them, the

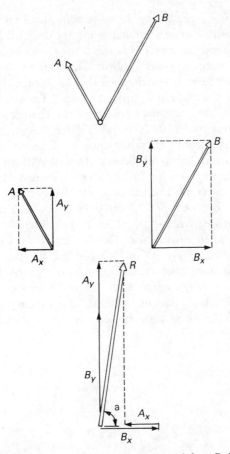

Fig. 1.18 *Resolution of force A into A_x and A_y, and force B into B_x and B_y. Algebraic addition of A_x and B_x horizontally, and of A_y and B_y vertically, to give the resultant, R.*

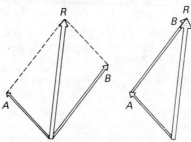

Fig. 1.19 *Determination of resultant, R, from forces A and B, by parallelogram of forces (left) and triangle of forces (right).*

length of the line representing the magnitude of the force. On the left, A and B are used to draw a parallelogram, the diagonal of which shows the magnitude and direction of R – this is the *parallelogram of forces*. On the right, the *triangle of forces* is shown. This does not directly represent the physical arrangement of the forces, since if force A is drawn first, force B is drawn starting at its tip, not its base. The resultant force R is represented by the line which completes the triangle, by joining the base of A to the tip of B. Both triangle and parallelogram give the same result, and it is a matter of convenience which one is used.

It follows from Newton's Second Law that if an object is not accelerating, there is no net force acting on it. Any forces which are acting on the object must be balanced out by other, equal and opposite forces. If the forces do not appear to balance, yet the object is not accelerating or decelerating, there must be a force (or forces) which have not been taken into account. This is illustrated in Fig. 1.20, which shows a complicated system of forces in use for the treatment of a fracture by balanced traction. Such problems are conveniently treated by drawing a *free body diagram* where all the forces are drawn as acting on a shapeless 'lump', floating in space, so that in any given direction the sum of all the forces must be zero. It can be seen that the forces

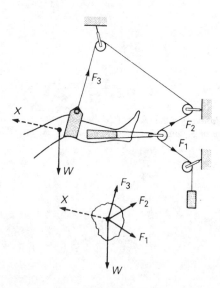

Fig. 1.20 *System of forces for balanced traction. Neglecting friction, F_1, F_2 and F_3 all equal the traction weight. W is the weight of the leg and X the reaction force on the patient. The equivalent free body diagram is shown below.*

from the traction system do not balance out – the missing force is the friction which prevents the patient from sliding down the bed. It can be found by a graphical method, which is an extension of the triangle of forces, or it can be calculated by resolving all the forces into vertical and horizontal components, calculating the 'missing' element of each, and then combining these components to give the magnitude and direction of the missing force.

Inertia and momentum

The term *inertia* is used to describe the resistance offered by a body to any attempt to set it in motion, or to stop it if it is already moving; it is a descriptive term, rather than a measured physical quantity. In the case of linear motion, it results from the mass of the object; in the case of rotational motion it describes how easy or difficult it is to rotate the object, which depends on how the mass is distributed about the center of gravity.

The *momentum* of a moving object is calculated by multiplying its velocity by its mass. A force, applied to the object, will cause it to change its velocity, and hence its momentum. Another way of expressing Newton's Second Law is to say that the force is equal to the rate of change of momentum.

Center of gravity

Although the mass of any object is distributed throughout every part of it, it is frequently convenient, as far as the effects of force are concerned, to imagine that the whole mass is concentrated at a single point, which is called the *center of gravity (C of G)*. For a regular shape, such as a cube, made of a uniform material, it is easy to see that the center of gravity must be at the geometric center. However, for irregular and changing shapes, such as the human body, it may be necessary to determine it by direct measurement. It is also possible to determine the center of gravity of every part of the body separately, and to find the center of gravity of the whole body by adding these together (by a method which is beyond the scope of this book).

It is frequently stated that the center of gravity of the body is just in front of the lumbosacral junction. This is approximately true for a person standing in the anatomical position, but any movement of the body will move the center of gravity. It is not even necessary for the center of gravity to remain within the body – the center of gravity of someone bending down to touch their toes will usually be outside the

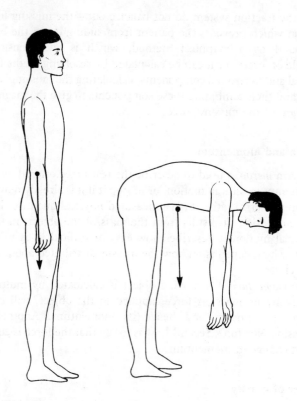

Fig. 1.21 *Center of gravity when standing and when bending.*

body, in front of the top of the thigh (Fig. 1.21). An interesting example of this is the technique used by skilled high-jumpers, who curve the body in such a way that although each part of the body in turn passes over the bar, the center of gravity may actually pass below it.

Moment of force

If a grown man wishes to play with a small girl on a see-saw, he will have to sit much closer to the center in order to balance the weight of the child (Fig. 1.22). The action which tends to unbalance the see-saw is the *moment of force*, which is calculated by multiplying the force by its perpendicular distance from the fulcrum or pivot point, this distance commonly being referred to as the *lever arm* or *moment arm*. The 'moment of force' may also be referred to as the 'torque', the 'turning

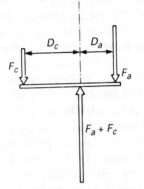

Fig. 1.22 *The adult on the see-saw will balance the child if the force* F_a *multiplied by the distance* D_a *equals* F_c *multiplied by* D_c.

moment', or simply the 'moment'. The formula for calculating the moment of force is:

$$M = F \times D$$

where M is the moment of force (in newton-meters, N·m), F is the force (in newtons, N), and D is the distance (in meters, m).

For the system to be in equilibrium, i.e. for the see-saw to balance, *the sum of the clockwise moments of force must equal the sum of the anticlockwise moments*. In Fig. 1.22, if the girl has a mass of 40 kg, she will exert a force due to gravity (F_c) of:

$$40\,\text{kg} \times 9.81\,\text{m/s}^2 = 392.4\,\text{N}$$

If her distance from the fulcrum (D_c) is 4 m, she exerts an anticlockwise moment of:

$$392.4\,\text{N} \times 4\,\text{m} = 1569.6\,\text{N·m}$$

A 70 kg man, producing a downward force (F_a) of:

$$70\,\text{kg} \times 9.81\,\text{m/s}^2 = 686.7\,\text{N}$$

will produce an opposing clockwise moment of 1569.6 N·m if he sits at a distance from the fulcrum (D_a) of:

$$1569.6 \text{ N·m}/686.7 \text{ N} = 2.29 \text{ m}$$

The definition of a moment of force refers to the perpendicular distance from the fulcrum. This is very important if opposing moments are produced by forces which act in different directions. Figure 1.23 illustrates the moments of force about the knee joint when standing with a bent leg. F_1, the weight of the body above this point, is acting at a perpendicular distance 'a' from the point of loadbearing. The quadriceps tendon is pulling at an oblique angle relative to the vertical, and the moment of force it provides is the product of the tension in the tendon, F_2, and the perpendicular distance 'b'. It will be noted that the presence of the patella increases the value of b and hence reduces the muscle force needed to produce a given moment of force. For equilibrium, the two moments ($F_1 \times a$) and ($F_2 \times b$) must be equal.

The measurement and interpretation of moments of force are essential for the full understanding of normal and pathological gait. Unfortunately, some confusion exists in the literature because a term like 'flexion moment' is often used without explaining whether it refers to an internal or external moment. Contraction of flexor muscles would generate an internal flexion moment, whereas an external flexion moment attempts to flex the joint, in opposition to the extensor muscles. To avoid such confusion, it is essential to make it clear

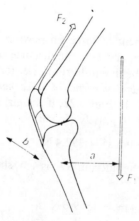

Fig. 1.23 *The moment due to gravity,* F_1 *multiplied by* a, *is opposed by contraction of the quadriceps which produces a moment* F_2 *multiplied by* b.

whether a moment is internal or external. Since the moments acting at a joint are usually estimated from external measurements, the author finds it most convenient to work with external moments. Although the internal moment must be equal in magnitude and opposite in direction to the external moment, it is often impossible to tell what structures are producing it. For example it could be generated by muscular contraction (concentric, isometric or eccentric), bone-on-bone forces, tension in the soft tissues, or moments transmitted from adjacent joints. The external moment of force is sometimes referred to as the 'reaction moment'.

Any object which is supported by the ground will remain stable so long as the line of force passing vertically downwards from its center of gravity remains within the area on the ground which is supporting it. Should the line of force stray outside this area, one of two things can happen – it may automatically correct itself, as happens with a self-righting lifeboat, or it may fall over, as will happen with a pencil balanced on its point. The former is a *stable equilibrium*, where a degree of imbalance produces 'restoring' moments which push the object back towards the balanced position. The latter is an *unstable equilibrium*, where the moments act to increase the imbalance. When walking at moderate speeds, there is an example of a further condition – a *dynamic equilibrium*, where from instant to instant the equilibrium is unstable, but before there has been time to fall, the area of support is moved and equilibrium is restored.

Linear motion

The *velocity* of a moving object is the rate at which its position changes, which usually means the distance it covers in a given time. It is, of course, similar to the everyday concept of speed, except that velocity is a vector, and thus has direction as well as magnitude. In measuring gait, the usual unit for velocity is meters per second, which can be abbreviated to either $m \cdot s^{-1}$ or m/s. Sometimes other units are used, such as meters per minute or kilometers per hour, but the SI units are to be preferred.

Acceleration is the rate at which velocity changes; the change may be in either magnitude or direction. If the velocity does not change, the acceleration is zero; a decrease in velocity may be known as negative acceleration, deceleration or retardation. If the velocity is measured in meters per second, the acceleration will be in meters per second per second, abbreviated to $m \cdot s^{-2}$ or m/s^2. The acceleration due to gravity has already been mentioned; it has a value of 9.81 m/s^2.

The relationships between velocity, acceleration and distance traveled are given by four equations:

$$v = u + at$$
$$s = \frac{1}{2}(ut + vt)$$
$$s = ut + \frac{1}{2}at^2$$
$$v^2 - u^2 = 2as$$

where u is the initial velocity (in meters per second, m/s)
v is the final velocity (in meters per second, m/s)
a is the acceleration (in meters per second per second, m/s^2)
t is the time (in seconds, s)
s is the distance traveled (in meters, m)

The worked example at the end of this section shows how one of these equations is used.

Circular motion

An object which is rotating has an *angular velocity*, and if the angular velocity changes there is an *angular acceleration* – it is easiest to think of a wheel rotating on its axle, and speeding up or slowing down. In walking, the leg has an angular velocity, and undergoes angular acceleration and retardation. An object which is not attached to an axle or fulcrum can also have an angular velocity, as it rotates about a point, such as its center of gravity. As with linear motion, angular acceleration will only occur if there is an application of force.

The detailed mathematics of angular velocity and angular acceleration are beyond the scope of this book, but it is worth saying a few words about the general concepts. Angular velocity is measured as angle turned per unit time, usually as degrees per second or radians per second. Angular acceleration is similarly expressed in degrees (or radians) per second per second. The radian is an obscure unit to nonmathematicians – it is the ratio, within the arc of a circle, of the length of the arc to the radius of the circle. There are 2π radians in a complete circle, giving the relationship: *$1 rad = 180°/\pi = 57.296°$.*

When a force applied to an object produces an angular acceleration, the acceleration does not depend solely on the size of the force and the mass of the object, as it does with linear motion. It also depends on the way in which the mass is distributed about the center of gravity – a property known as the *moment of inertia*. An object with the mass concentrated around the outside, such as a flywheel, has a much higher moment of inertia than one with the mass concentrated around the center, such as a cannon ball. If a flywheel and a cannon ball have the same mass, and are spinning with the same angular velocity, the flywheel will prove much more difficult to stop than the cannon ball.

Kinetics and kinematics

The terms kinetics and kinematics are commonly used in gait analysis, and they deserve some explanation. *Kinetics* is the study of force, moments, mass and acceleration, but without any detailed knowledge of the position and orientation of the objects involved. For example, an instrument known as a force platform can be used to measure the force below the foot during walking, but it gives no information on the position of the limb or the angle of the joints. *Kinematics* describes motion, but without reference to the forces involved. An example of a kinematic instrument is a camera, which can be used to observe the motion of the trunk and the limbs during walking, but which gives no information on the forces involved. It is fairly obvious that for an adequate quantitative description of an activity such as walking, both kinetic and kinematic data are needed.

Work, energy and power

One of the remarkable features of normal gait is how energy is conserved by means of a number of optimizations. However, abnormal gait patterns may result in excessive fatigue, and the measurement of energy consumption during walking can be an important element in scientific gait analysis.

There is a subtle difference in viewpoint between the physical scientist and the biologist as far as work, energy and power are concerned. To the physical scientist, *work* is done when a force moves an object over a distance. It is calculated as the product of the force and the distance – if a force of two newtons moves an object three meters, the work done is: $2N \times 3m = 6J$ *(joules)*.

The *joule* could also be called a newton-meter, but this would cause confusion with the identically named unit used to measure moments of force. *Energy* is the capacity to do work, and is also measured in joules. It exists in two basic forms – *potential*, or stored energy, and *kinetic*, or movement energy. In walking, there are alternating transfers between potential and kinetic energy, which will be described in Chapter 2. *Power* is the rate at which work is done – a rate of one joule per second is a *watt*, which is familiar to users of electrical appliances.

The reason biologists regard these matters slightly differently from physical scientists is that muscles can use energy without shortening, in other words without doing physical work. The potential energy stored in the muscles, in the form of ATP, is converted to mechanical energy in response to the muscle action potential. This energy is still used, even in an eccentric contraction, where the muscle actually gets longer while developing a force, which in physical terms is negative work. In other words, while everyone agrees that walking uphill involves the doing of work, the physicist might expect someone walking downhill to gain energy, whereas in reality the muscles are still activated, and metabolic energy is still expended. Even if the muscle shortens as it contracts, in a concentric contraction, the conversion of metabolic energy to mechanical energy is relatively inefficient, with a typical efficiency of around 25 per cent. The old unit for measuring metabolic energy was the Calorie (the capital C indicating 1000 calories or 1 kilocalorie); the conversion factor is 4200J or 4.2 kJ per Calorie (see Appendix 2).

The calculation of the mechanical power generated at joints has become an important part of the biomechanical study of gait. In a rotary movement, when a joint flexes or extends, the power is calculated as the product of the moment of force and the angular velocity, omega (ω):

$$P \; (watts) = M \; (newton\text{-}meters) \; x \; \omega \; (radians \; per \; second)$$

Worked example

As an example of the biomechanical principles outlined above, consider the mechanics involved when 'Wonder Woman' (one of the fictional 'superheroines') jumps from the ground to the top of a building 10 m high. In order to perform this superhuman feat, she must leave the ground with sufficient velocity to reach the top of the building, despite the deceleration produced by gravity. The distance to be travelled (s) is 10 m; the initial velocity (u) is unknown; the final velocity (v) is zero, and the acceleration (a) is $-9.81 \, \text{m/s}^2$, which is the acceleration due to gravity, negative because it opposes the movement. The substitution of these values in the equation:

$$v^2 - u^2 = 2as$$

gives the initial velocity (u) as 14.0 m/s.

In order to achieve the necessary velocity, Wonder Woman must accelerate from being stationary in the crouched position to a velocity of 14.0 m/s in the fully stretched position, as the feet leave the ground. The center of gravity of the body, in making this jump, moves through a vertical distance (s) of 0.7 m. The initial velocity (u) is zero, the final velocity (v) is 14.0 m/s. Substituting these values in the equation used above gives the average acceleration (a) as being 140 m/s², or about 14 times the acceleration due to gravity, and the duration of the acceleration as 0.1 seconds.

To achieve this acceleration requires a force acting vertically upwards on the center of gravity. The force is given by the equation:

$$F = m \times a$$

which follows from Newton's Second Law, where the force (F) is equal to the body mass (m) multiplied by the acceleration (a). If her body mass is 50 kg, the force is: 50 kg × 140 m/s² = 7000 N or 7 kN.

Neglecting the small contribution made by swinging the arms, this force is applied to the center of gravity of the body by extension of the two knees. When the knee is flexed to a right angle, about halfway through the acceleration, the center of gravity is about 0.4 m behind the knee joint, whereas the quadriceps tendon is only 0.06 m in front of it. To apply an upward force of 7 kN to the center of gravity, the moment of force required to be generated by each knee is: 7 kN × 0.4 m × ½ = 1.4 kN·m.

Since the sum of the clockwise and anticlockwise moments must be equal, the quadriceps must generate a force of: 1.4 kN-m/0.06 m = 23.3 kN, or about 48 times body weight. Needless to say, only someone like Wonder Woman can generate quite such large forces in the quadriceps tendon, mere mortals having to settle for about a fifth as much!

The knee extends from about 160° of flexion to 0° in 0.1 seconds – its angular velocity is thus 1600° per second, or close to 30 radians per second. The power generated is the product of the moment of force and the angular velocity, which for the moment calculated above is: 1.4 kN·m × 30 rad/s = 42 kW. This contrasts with a power generation at the knee in normal walking of 100–200 watts.

2
Normal Gait

In order to understand abnormal gait, it is first necessary to study normal gait, since this provides the standard against which the gait of a patient can be judged. However, there are two pitfalls which need to be borne in mind when using this approach. Firstly, the term 'normal' covers both sexes, a wide range of ages, and an even wider range of extremes of body geometry, so that an appropriate standard needs to be chosen for the individual who is being studied. If results from an elderly female patient are compared with normal data obtained from physically fit young men, there will undoubtedly be large differences, whereas comparison with normal data from other elderly women may show the patient's gait to be well within normal limits appropriate to her sex and age. The second pitfall is that even though a patient's gait differs in some way from normal, it does not follow that this is in some way undesirable, or that efforts should be made to turn it into a normal gait. Many gait abnormalities are a compensation for some problem experienced by the patient, and although abnormal, they are nonetheless useful.

Having said that, it is very important to understand normal gait, and the terminology which is used to describe it, before going on to look at pathological gait. The chapter starts with a very brief historical review, then gives an overview of the gait cycle, before going on to study in detail how the different parts of the locomotor system are used in walking.

Walking and gait

As walking is such a familiar activity, it is difficult to define it without sounding pompous. However, it would be remiss not to attempt a definition. Normal human walking and running can be defined as 'a method of locomotion involving the use of the two legs, alternately, to provide both support and propulsion.' In order to exclude running, we must add '. . . at least one foot being in contact with the ground at all times.' Unfortunately, this definition excludes some forms of pathological gait which are generally regarded as being forms of walking, such as the 'swing-through gait' (see Fig. 3.20), in which there is an alternate use of two crutches and either one or two legs. It is probably both pointless and unreasonable to attempt to arrive at a definition of walking which will apply to all cases – at least in a single sentence!

Gait is no easier to define than walking, many dictionaries regarding it as a word primarily for use in connection with horses! Most people, including the author, tend to use the words gait and walking interchangeably. However, there is a difference: the word gait describes the manner or style of walking, rather than the walking process itself. It thus makes more sense to talk about a difference in gait between two individuals than about a difference in walking.

History

The history of gait analysis has shown a steady progression from early descriptive studies, through increasingly sophisticated methods of measurement, to mathematical analysis and mathematical modeling. Only a brief account of the development of the discipline will be given here. Good reviews of the early years of gait analysis have been given by Garrison (1929), Bresler and Frankel (1950) and Steindler (1953). The more recent history is particularly difficult, as several hundred papers have been published on the subject in the last two decades. For a fairly complete bibliography, the reader is referred to Vaughan et al. (1987). This section will cite just a few publications which the author thinks are particularly significant.

Descriptive studies

Gait has undoubtedly been observed ever since man evolved, but the systematic study of the subject appears to date from the Renaissance, when Leonardo da Vinci, Galileo and Newton all gave useful descriptions of walking. The earliest account using a truly scientific approach was in the classic 'De Motu Animalum', published in 1682 by Borelli, who worked in Italy and was a student of Galileo. Borelli measured the center of gravity of the body, and described how balance is maintained in walking by constant forward movement of the supporting area provided by the feet.

The Weber brothers in Germany gave the first clear description of the gait cycle in 1836. They made accurate measurements of the timing of gait and of the pendulum-like swinging of the leg of a cadaver.

Kinematics

Two pioneers of kinematic measurement worked on opposite sides of the Atlantic in the 1870s. Marey, working in Paris, published a study of human limb movements in 1873. He made multiple photographic exposures, on a single plate, of a subject who was dressed in black, except for brightly illuminated stripes on the limbs. He also investigated the path of the center of gravity of the body, and the pressure beneath the foot. Eadweard Muybridge (born Edward Muggeridge) became famous in California in 1878, by demonstrating that, when a horse is trotting, there are times when it has all its feet off the ground at once. The measurements were made using 24 cameras, triggered in quick succession as the horse ran into thin wires stretched across the track. In the next few years, Muybridge made a further series of studies, of naked human beings walking, running and performing a surprising variety of other tasks!

The most serious application of the science of mechanics to human gait during the nineteenth century was the publication, in Germany in 1895, of 'Der Gang des Menschen', by Braune and Fischer. They used a technique similar to Marey's, but using fluorescent strip-lights on the limbs instead of white stripes. The resulting photographs were used to determine the three-dimensional trajectories, velocities and accelerations of the body segments. Knowing the masses and accelerations of the body segments, they were then able to estimate the forces involved at all stages during the walking cycle.

Further valuable work on the dynamics of locomotion was done by Bernstein, working in Moscow in the 1930s. He developed a variety of photographic techniques for kinematic measurement, and studied over 150 subjects. Particular attention was paid to the center of gravity of the individual limb segments and of the body as a whole.

Force platforms

Further progress followed the development of the force platform (also called the forceplate). This instrument has contributed greatly to the scientific study of gait, and is now standard equipment in most gait laboratories. It measures the direction and magnitude of the ground reaction force beneath the foot. An early design was described by Amar in 1924, and an improved one by Elftman in 1938. Both were purely mechanical, the force applied to the plate causing the movement of a pointer. In Elftman's design the pointers were photographed by a high-speed cine camera.

Muscle activity

For a full understanding of normal gait, it is necessary to know which muscles are active during the different parts of the gait cycle. The role of the muscles was studied by Scherb, in Switzerland during the 1940s, initially by palpating the muscles as his subject walked on a treadmill, and later by the use of electromyography (EMG).

Further advances in the understanding of muscle activity and many other aspects of normal gait were made during the 1940s and 1950s by a very active group working in the University of California at San Francisco and Berkeley, notable among whom was Verne Inman. Some members of this group later went on to write 'Human Walking' (Inman et al., 1981), published just after Inman died, and which to many people is the definitive textbook on normal gait.

Mechanical analysis

A major contribution to the mechanical analysis of walking, also from the Californian group, was made by Bresler and Frankel (1950). They performed free-body calculations for the hip, knee and ankle joints, allowing for ground reaction forces, the effects of gravity on the limb segments and the inertial forces. The analytical techniques described by these workers have formed the basis of many current methods of joint modeling and analysis.

An important paper describing the possible mechanisms which the body uses to minimize energy consumption in walking, again from California, was published by Saunders et al. (1953). Further important work on energy consumption, and in particular the energy transfers between the body segments in walking, was published by Cavagna and Margaria (1966), working in Italy.

By 1960, research began to concentrate on the variability of walking, the development of gait in children, and the deterioration of gait in old age. Patricia Murray, working in Milwaukee, Wisconsin, published a series of papers on these subjects, including a detailed review (Murray, 1967).

Mathematical modeling

Once the motions of the body segments and the actions of the different muscles had been examined and documented, attention passed to the forces generated across the joints. Although limited calculations of this type had been made previously, the study by Paul (1965) was the first detailed analysis of hip joint forces during walking. A subsequent paper by the same author also included an analysis of the forces in the knee (Paul, 1966). Since then there have been many mathematical studies of force generation and transmission across the hip, knee and ankle.

The 1970s and 1980s saw great improvements in methods of measurement. The development of more convenient kinematic systems, based on electronics rather than photography, meant that results could be produced in minutes rather than days. Reliable force platforms with a high frequency response became available, as well as convenient and reliable EMG systems. The availability of high quality three-dimensional data on the kinetics and kinematics of walking, and the ease of access to powerful computers, made it possible to develop increasingly sophisticated mathematical models, which could calculate the muscle, ligament and joint contact forces in the lower limb.

Clinical application

From the earliest days, it has been the hope of most of those working in this field that gait measurements would be found useful in the management of patients with walking disorders. Many of the early workers made studies of people who walked with abnormal gait patterns, and some (notably Amar, Scherb and the Californian group) attempted to use the results for the benefit of individual patients. However, the results were not particularly impressive.

Since 1960, there has been a more serious attempt to take gait analysis out of the research laboratory and into the clinic. With the improvements in measurement and analytical techniques, the major limitation now is not the ability to produce high quality data, but knowing how best to use these data for the benefit of patients. It is fair to say that far more progress has been made in scientific gait analysis, particularly as applied to normal subjects, than in the application of these techniques for the benefit of those with gait disorders. However, a small number of clinicians, most of them orthopaedic surgeons, have shown the value of gait analysis in patient care.

As well as a gradual increase in the clinical use of scientific gait analysis, there has also been a growing interest in the use of observational or visual gait analysis. This has become much easier to perform since video cameras and video cassette recorders have become widely available.

Terminology used in gait analysis

The *gait cycle* is defined as the time interval between two successive occurrences of one of the repetitive events of walking. Although any event could be chosen to define the gait cycle, it is usually convenient to use the heel contact of one foot. If it is decided to start with the right foot, as shown in Fig. 2.1, then the cycle will last until the next heel contact by the right foot. The left foot, of course, goes through exactly the same series of events as the right, but displaced in time by half a cycle. Each leg, in turn, has a *swing phase*, when it moves forwards through the air, and a *stance phase*, when the foot is on the ground, and the body passes over the top of it.

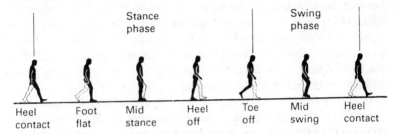

Fig. 2.1 *Positions of the legs during a single gait cycle from right heel contact to right heel contact.*

The following major events are used to divide the gait cycle into convenient periods for the purposes of description:

1. Heel contact
2. Foot flat
3. Mid stance } Stance phase
4. Heel off
5. Toe off
6. Mid swing } Swing phase
(1. Heel contact)

The stance phase, which is also called the 'support phase' or 'contact phase', lasts from heel contact to toe off. The swing phase lasts from toe off to the next heel contact. The duration of the complete gait cycle is known as the *stride time*; it is divided into the *stance time* and the *swing time*. The naming of the events of the gait cycle varies from one publication to another; the present text has largely followed the recommendations of Winter (1987). The term 'heel contact' has been used as a compromise between the more usual 'heelstrike' and the general term 'initial contact', which is confusing when it is used to describe the end of the cycle. Alternative terminology will be given at the appropriate points in the detailed description. Wall et al. (1987) pointed out that the usual terminology may not be adequate to describe some severely pathological gaits.

Gait cycle timing

Figure 2.2 shows the timings of heel contact and toe off for both feet during a single gait cycle. Heel contact by the right foot occurs while the left foot is still on the ground, and there is a period of *double support* between heel contact on the right and toe off on the left. During the swing phase on the left side, only the right foot is on the ground, giving a period of *right single support*, which ends with heel contact by the left foot. There is then another period of double support, until toe off on the right side. *Left single support* corresponds to the right swing phase, and the cycle ends with the next heel contact on the right.

In each gait cycle, there are thus two periods of double support and two periods of single support. The stance phase usually lasts for about 60 per cent of the cycle, the swing phase for about 40 per cent, and each period of double support for about 10 per cent. However, this varies with the speed of walking, the swing phase becoming proportionately longer, and the stance phase and double support phases shorter, as the speed increases (Murray, 1967). The final disappearance of the double support phase marks the transition from walking to running. Between

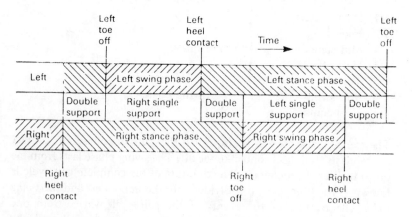

Fig. 2.2 *Timing of single and double support during a single gait cycle from right heel contact to right heel contact.*

successive steps in running there is a *flight phase*, when neither foot is on the ground.

In each double support phase, one foot is forward, having just landed on the ground, and the other one is backward, being just about to leave the ground. The foot which is forward is sometimes referred to as being in 'braking double support', 'loading response' or 'weight acceptance'. The backward leg is in 'thrusting double support', 'pre-swing', 'terminal stance' or 'weight release'.

Foot placement

The terms used to describe the placement of the feet on the ground are shown in Fig. 2.3. The *stride length* is the distance between two successive placements of the same foot. It consists of two *step lengths*, left and right, each of which is the distance by which the named foot moves forward in front of the other one. In pathological gait, it is perfectly possible for the two step lengths to be different. If the left foot is moved forward to take a step, and the right one is brought up beside it, rather than in front of it, the right step length will be zero. It is even possible for the step length on one side to be negative, if that foot never catches up with the other one. However, the stride length measured between successive positions of the left foot must always be the same as that measured from the right foot, unless the subject is walking around a curve.

This definition of a 'stride', consisting of one 'step' by each foot, breaks down in some pathological gaits, in which one foot makes a

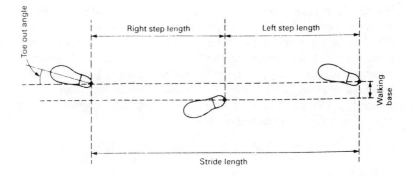

Fig. 2.3 *Terms used to describe foot placement on the ground.*

series of 'hopping' movements while the other is in the air (Wall et al., 1987). There is no satisfactory nomenclature to deal with this situation.

The *walking base* (also known as the 'stride width') is the side-to-side distance between the line of the two feet, usually measured at the mid point of the heel, but sometimes at the center of the ankle joint. The preferred units for stride length and step length are meters, and for the walking base millimeters.

The *toe out* (or, less commonly, *toe in*) is the angle in degrees between the direction of progression and a reference line on the sole of the foot. The reference line varies from one study to another; it may be defined anatomically, but is commonly the midline of the foot, as judged by eye.

Cadence and velocity

The *cadence* is the number of steps taken in a given time, the usual units being steps per minute. In most other types of scientific measurement, complete cycles are counted, but as there are two steps in a single gait cycle, the cadence is a measure of half-cycles.

The *velocity* of walking is the distance covered by the whole body in a given time, in a particular direction. It should be measured in meters per second. The instantaneous velocity varies from moment to moment during the walking cycle, but the average velocity is the product of the cadence and the stride length. The cadence, in steps per minute, corresponds to half-strides per 60 seconds, or full strides per 120 seconds. The velocity can thus be calculated from cadence and stride length using the formula:

$$\text{velocity (m/s)} = \text{stride length (m)} \times \text{cadence (steps/min)} / 120$$

It is clearly illogical to measure cadence in steps per minute and velocity in meters per second. However, these are the units most commonly found in the gait analysis literature, and they have been adopted for the present text. A few people working in the field measure cadence in steps per second, which is more consistent, and it is hoped that this usage will eventually replace the earlier one. Another way to avoid this difficulty is to replace cadence by a quantity which is inversely related to it – the 'cycle time' in seconds:

$$\text{cycle time (s)} = 120 / \text{cadence (steps/min)}$$

The gait cycle

Outline

The purpose of the present section is to provide the reader with an overview of the gait cycle, to make the detailed description in the following section easier to understand. The cycle is illustrated in Figs 2.4 to 2.7 and 2.9 to 2.14, all of which are taken from a single walk by a 20-year-old normal female, walking with a cadence of 109 steps/minute, a stride length of 1.45 m and a velocity of 1.32 m/s. The results from this subject do not correspond to 'average' values for some of the parameters, because of the normal variability between individuals, although they are all within the normal range. The measurements were all made in the plane of progression, which is the vertical plane most closely corresponding to the direction of the walk.

Figure 2.4 shows the successive positions of one leg at 40 ms intervals, measured over a little more than one gait cycle. Figure 2.5 shows the corresponding angles at the hip, knee and ankle joints. Figure 2.6 shows a 'butterfly diagram' made up of successive representations, at 20 ms intervals, of the ground reaction force vector. Rose (1985) has found this way of displaying force platform data to be useful in a clinical setting. The vectors move across the diagram from left to right, the magnitude of each vector being shown by the length of the line. The data were obtained using a Vicon television/computer system and a Kistler force platform, both of which are described in more detail in Chapter 4.

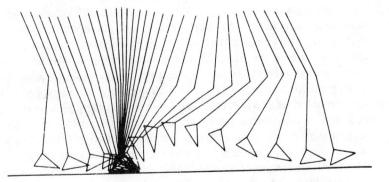

Fig. 2.4 *Position of right leg at 40 ms intervals during slightly more than one gait cycle.*

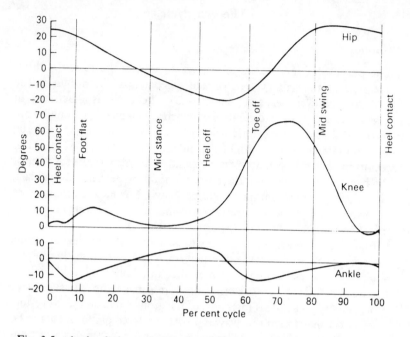

Fig. 2.5 *Angles during a single gait cycle of hip (flexion positive), knee (flexion positive) and ankle (dorsiflexion positive).*

Upper body: The upper body moves forwards throughout the gait cycle. Its velocity varies a little, being fastest during the double support phases and slowest in the middle of the stance and swing phases. The trunk twists about a vertical axis, the pelvis and shoulder girdle rotating in opposite directions. The arms swing out of phase with the legs, so that as the left leg and the left side of the pelvis move forwards, so do the right arm and the right side of the shoulder girdle. Murray (1967) found average total excursions of 7 degrees for the shoulder girdle and 12 degrees for the pelvis, in adult males walking at free speed. The whole trunk rises and falls twice during the cycle, through a total range of about 50 mm, being lowest during double support, and highest around mid stance or mid swing. An approximation to this vertical motion can be seen in the position of the hip joint marker in Fig. 2.5. The trunk also moves from side to side, once in each cycle, being over each leg during its stance phase. The total range of movement is again about 50 mm. The pelvis, as well as twisting about a vertical axis, also tips slightly, both backwards and forwards and from side to side.

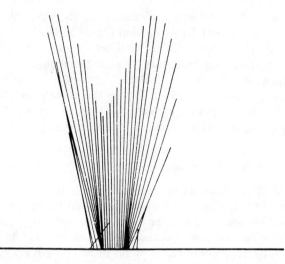

Fig. 2.6 *'Butterfly diagram' of ground reaction force vector at 20 ms intervals. Progression is from left to right.*

Hip: The hip flexes and extends once during the cycle (Fig. 2.5). The limit of flexion is reached about the middle of the swing phase, and the thigh is then kept flexed until the beginning of the stance phase. The peak extension is reached before the end of the stance phase, after which the hip begins to flex again. The hip angle shown in Fig. 2.5, and in similar diagrams throughout the book, is actually the angle between the thigh and the vertical, and not the true hip angle, which is between the thigh and the pelvis. However, movement of the pelvis in the sagittal plane is usually small, and the traces are very similar.

Knee: The knee shows two flexion and two extension peaks during each gait cycle. It is fully extended before heel contact, flexes early in the stance phase ('stance phase flexion'), extends again around mid stance, then starts flexing again, reaching a peak early in the swing phase ('swing phase flexion'). It extends again prior to the next heel contact.

Ankle and foot: The ankle is usually within a few degrees of the neutral position at the time of heel contact, with the heel slightly inverted and the foot slightly supinated. After heel contact, the ankle plantarflexes, bringing the foot down flat on the ground and moving it into pronation. As the stance phase progresses, the tibia moves forwards over the foot, and the ankle angle changes from being plantarflexed to being dorsiflexed. The tibia also rotates externally, which moves the foot, by means of the subtalar joint, from pronation into supination (Inman et al., 1981). At heel off, the hindfoot lifts and the ankle angle again changes, moving back into plantarflexion until toe off. While the forefoot is on the ground and the heel is off it, the heel is inverted and the foot remains supinated. During the swing phase, the ankle moves back into dorsiflexion until the forefoot has cleared the ground, after which the neutral position is adopted prior to the next heel contact.

The gait cycle in detail

Each of the following sections begins with some general remarks about the events leading up to a particular point in the gait cycle, and then describes what is happening in the upper body, hips, knees, ankles and feet, with particular reference to the activity of the muscles. More detailed descriptions of the events of normal gait are given by Murray (1967) and by Inman et al. (1981).

Heel contact

1. *General:* Heel contact (Fig. 2.7) is the beginning of the stance phase. It is frequently called 'heelstrike', since there is often a distinct

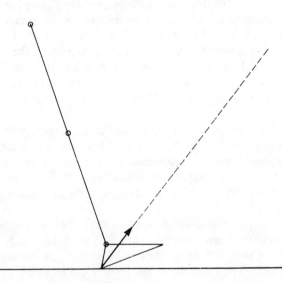

Fig. 2.7 *Position of limb and ground reaction force vector at heel contact (during heelstrike transient).*

impact between the heel and the ground, known as the 'heelstrike transient'. In pathological gait, where some other part of the foot may contact the ground first, the term 'initial contact' is to be preferred. There is considerable variation between individuals as to how much force is applied to the ground at heel contact, some people 'gliding' the foot onto the ground, and others 'digging' it in. Figure 2.8 shows a trace of the vertical component of the ground reaction force measured from an individual with a marked heelstrike (Collins and Whittle, 1989). There have been suggestions that the transient forces in the joints resulting from the heelstrike may lead to degenerative arthritis. The heelstrike transient is fairly short, typically lasting between 10 and 20 ms, and it can only be observed using measuring equipment with a fast enough response time. The direction of the ground reaction force changes from upwards and forwards during the heelstrike transient (Fig. 2.7), to upwards and backwards immediately afterwards (Fig. 2.9). This change in direction can also be seen in Fig. 2.6, where the initial vector appears to point 'the wrong way'.

2. *Upper body:* The trunk is about half a stride length behind the level of the foot at the time of heel contact. It is at its lowest vertical position, about 25 mm below its average level for the whole cycle, and its instantaneous forward velocity is at its highest, perhaps 10 per cent above its average velocity. In the side-to-side direction, the trunk is

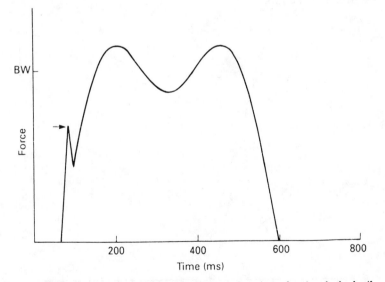

Fig. 2.8   *Plot of vertical ground reaction force against time, showing the heelstrike transient (arrowed).* BW = *body weight.*

crossing the midline of its range of travel, moving towards the side of the leg which has just made contact. The trunk is twisted, the left shoulder and the right side of the pelvis being at their furthest forwards. The left arm is at its most advanced when the right foot undergoes heel contact. The amount of arm swinging varies greatly from one person to another, and it also increases with the speed of walking. At the time of heel contact, Murray (1967) found that the mean elbow flexion was 8 degrees and the mean shoulder flexion 45 degrees.

3. *Hip:* The attitude of the legs at the time of heel contact is shown in Fig. 2.7. The maximum flexion of the hip (generally around 30 degrees) is reached around the middle of the swing phase, after which it extends again only slightly before heel contact. The hip then extends more rapidly, through contraction of the hip extensors, the hamstrings being active during the latter part of the swing phase, and gluteus maximus starting to contract around heel contact.

4. *Knee:* During the early swing phase, the knee flexes to 60 or 70 degrees. It then extends again, becoming more or less straight just before heel contact (see Figs 2.5 and 2.7). Except in very slow walking, the knee flexors contract eccentrically at the end of the extension, to act as a braking mechanism to prevent hyperextension. The hamstrings usually provide this function, in addition to their role in initiating extension of the hip.

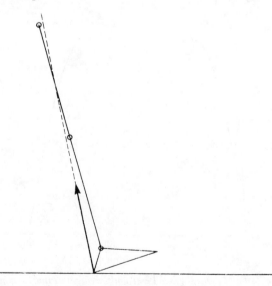

Fig. 2.9   *Position of limb and ground reaction force vector 20 ms after heel contact.*

5. *Ankle and foot:* The ankle is close to its neutral position during the latter part of the swing phase. If the foot is moving backwards as the heel contacts the ground, the ground reaction to the heelstrike transient is directed forwards (see Fig. 2.7). The direction of the force vector changes to that shown in Fig. 2.9 within 10–20 ms, and an external plantarflexing moment is produced. The foot would come down much too quickly, in a 'foot-slap', if this moment were not resisted by the anterior tibial muscles, which contract eccentrically to lower the foot gently to the ground. The heel is usually slightly inverted and the foot slightly supinated at the time of heel contact, and most people show a wear pattern on the lateral side of the heel of the shoe.

Foot flat

1. *General:* After heel contact, the rest of the foot comes down onto the ground at foot flat (see Fig. 2.10), which generally occurs at around 8 per cent of the gait cycle, just before toe off on the other side. During the interval between heel contact and foot flat, the ground reaction force increases rapidly in magnitude, its direction being upwards and backwards. This is generally known as the *weight acceptance* phase of gait, although the term 'load acceptance' is to be preferred, since inertia is involved as well as weight. Although the ground reaction force has a backward directed component, this may be cancelled out by the

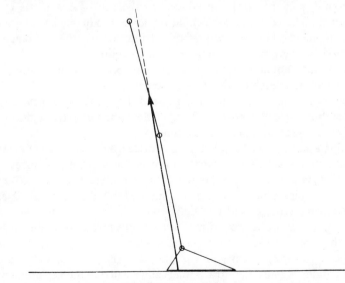

Fig. 2.10   *Position of limb and ground reaction force vector at foot flat.*

forward directed component of the ground reaction force from the other foot, while it is still on the ground.

2. *Upper body:* The shoulder and arms, having reached their most advanced position on the contralateral side, are now moving back again. Similarly, the pelvis on the side of the stance phase leg now starts to twist back towards the neutral position. The trunk reaches its lowest position between heel contact and foot flat, and starts to lift upwards, slowing its forward motion slightly as it does so.

3. *Hip:* The hip is 20 to 25 degrees flexed at the time of foot flat, and continuing to extend, by contraction of the gluteus maximus and the hamstrings. This extension is opposed by an external flexion moment. In Figs 2.9 and 2.10, the projected ground reaction force vector can be seen to pass in front of the hip. However, this only gives a rough guide to the external moments being generated at the hip. The external force applied to any joint differs from the ground reaction force by the mass and inertia of however much of the leg is between that joint and the ground. In the case of the hip, the whole leg is involved, and the errors can be considerable.

4. *Knee:* The knee flexes following heel contact, and in doing so it acts as a spring, preventing the vertical force from building up too rapidly (Perry, 1974). The knee moment can be estimated from the projected force vector in Fig. 2.9, rather more reliably than it can for the hip, since only the shank and foot are between the knee joint and the ground, and the errors introduced by their mass and inertia are fairly small. The vector passes behind the knee, producing an external flexion moment. This is opposed by an internal extension moment generated by the quadriceps muscles, which contract eccentrically to permit a small amount of controlled flexion. Knee flexion continues beyond foot flat, reaching the peak of stance phase flexion at 15 to 20 per cent of the gait cycle. The magnitude of the stance phase flexion is variable, but is usually between 10 and 20 degrees. The external flexion moment also reaches a peak at this time.

5. *Ankle and foot:* The stage of the gait cycle from heel contact to foot flat, sometimes called the *initial rocker* or 'heel pivot', involves a plantarflexion at the ankle of around 15 degrees, which is achieved by an external plantarflexion moment, opposed by the anterior tibial muscles contracting eccentrically. As soon as the foot is flat on the ground, the line of the ground reaction force begins to move forwards along the foot (see Fig. 2.10), causing the moment to become smaller and then to reverse. The movement into plantarflexion is accompanied by pronation, because the tibia is internally rotated at the beginning of the stance phase (Inman et al., 1981).

Mid stance

1. *General:* The term 'mid stance' may be used either to describe the period of time between foot flat and heel off, or to define a particular event occurring during that period of time. It is used here in the latter sense, to mean the time at which the swing phase leg passes the stance phase leg, and the two feet are side by side. Mid stance (see Fig. 2.11), defined in this way, is close to the midpoint of the stance phase, and is about 30 per cent of the way through the whole gait cycle. However, it is not necessarily close to the midpoint of the period between foot flat and heel off, since both of these events vary, both from one person to another and with the speed of walking. Another definition of 'mid stance' is the time at which the greater trochanter of the femur is vertically above the midpoint of the foot, in the sagittal plane.

2. *Upper body:* The period from foot flat to mid stance sees the trunk climbing to its highest point, about 25 mm above the mean level, and slowing its forward velocity, as the kinetic energy of forward motion is converted to the potential energy of height. The side-to-side motion of the trunk also reaches its peak at mid stance, the trunk being displaced about 25 mm from its central position, towards the side of the stance leg. Like the feet, the arms are side by side at mid stance, as each follows the motion of the opposite leg. The twisting of the trunk has now disappeared, as both the shoulder girdle and pelvis pass through neutral before twisting the other way.

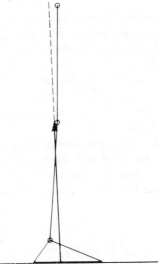

Fig. 2.11   *Position of limb and ground reaction force vector at mid stance.*

3. *Hip:* At mid stance, the hip is more than half way through its movement into extension, with a typical angle close to zero (see Fig. 2.11). The external flexion moment, and the opposing contraction of the extensor muscles, declines and disappears during the middle of the stance phase. It is replaced by a moment in the opposite direction, an external extension moment, which is opposed by an internal moment generated by the hip flexors, psoas major and iliacus, contracting eccentrically. The other significant muscle activity about the hip joint takes place in the coronal plane. As soon as the opposite foot has left the ground, the pelvis is supported only by the stance phase hip. Although dipping down slightly on the side of the swinging leg, its position is maintained by contraction of the hip abductors, especially gluteus medius.

4. *Knee:* By mid stance, the knee has started to extend again. The external flexion moment reduces in magnitude, but is still present, since the ground reaction force is still behind the knee. As the quadriceps muscles have finished contracting by this time, there is, at first glance, nothing to generate an internal moment to resist flexion at the knee. However, such a moment is generated by the dual effects of the contraction of the soleus and the forward motion of the upper body. If the ankle joint were totally free, the forward motion would simply dorsiflex the ankle, but contraction of the soleus slows down the forward motion of the tibia, and as the femur continues to move forwards, the knee extends.

5. *Ankle and foot:* The period from foot flat to heel off, sometimes called the 'mid-stance rocker', is characterized by forward rotation of the tibia about the ankle joint, from about 15 degrees of plantarflexion to about 10 degrees of dorsiflexion. At mid stance the ankle is generally between the neutral position and 5 degrees of dorsiflexion (see Fig. 2.11). The ground reaction force vector moves forward along the foot from the time of foot flat onwards, the initial external plantarflexion moment being replaced by an external dorsiflexion moment of increasing magnitude. As mentioned above, this is resisted by contraction of the soleus, which prevents the tibia from moving forward too quickly. As the stance phase progresses, the tibia rotates externally, and the subtalar joint causes the foot to move out of pronation.

Heel off

1. *General:* Heel off (see Fig. 2.12), also called 'heel rise', is the time at which the heel begins to lift from the walking surface. Its timing varies considerably, both from one individual to another and with the speed of

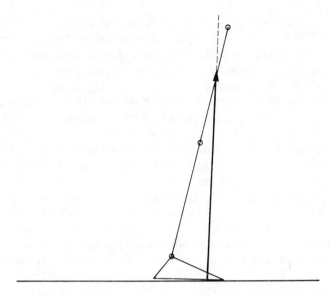

Fig. 2.12   *Position of limb and ground reaction force vector at heel off.*

walking. The descriptions which follow are based on a typical heel off at 40 per cent of the gait cycle, although in the normal subject shown in Figs 2.4 to 2.7 and 2.9 to 2.14, heel off was a little later, at 45 per cent. Heel off normally occurs before heel contact by the other leg, which occurs at around 50 per cent.

2. *Upper body:* Once mid stance has passed, the trunk begins to lose vertical height, on its way down to its lowest point in the double support phase. The lateral displacement over the support leg also begins to reduce, in preparation for the transfer of weight back to the other leg. As the hip extends and the stance leg moves backwards, the pelvis twists backwards with it, and the arms and shoulder girdle on that side move forwards.

3. *Hip:* The hip continues to extend, reaching an angle of between 10 and 15 degrees of extension by the time of heel off. The external extension moment continues to act, and the hip flexors, mainly psoas and iliacus, contract eccentrically to resist this. Between heel off and toe off, the hip reaches a peak of extension and starts to flex again. The activity of the hip abductors in the coronal plane is still required, until heel contact by the other foot.

4. *Knee:* The knee has an extension peak close to the time of heel off, the angle at this time being between zero and a few degrees of flexion. Between mid stance and heel off, the ground reaction force vector moves

in front of the knee joint, so producing an external extension (or hyperextension) moment. The role of the soleus in opposing the external flexion moment around mid stance was described previously. Once the direction of the external moment changes, so that it attempts to extend the knee, the gastrocnemius comes into play. This muscle augments the action of the soleus as far as the ankle joint is concerned, but it also acts as a flexor at the knee, and so prevents hyperextension.

5. *Ankle and foot:* The peak of ankle dorsiflexion is reached around the time of heel off. The ankle angle at this time is usually between 15 and 20 degrees, but was less in the normal subject used for illustration (see Fig. 2.5). As soon as the heel leaves the ground, the knee begins to flex and the ankle to plantarflex. As the force vector moves forwards along the foot, the external dorsiflexion moment increases, and first the soleus and then the soleus and gastrocnemius together contract to oppose it, the peak activity by these muscles coinciding with the peak external moment, just after heel off. As the heel rises it also inverts, and the foot supinates.

Toe off

1. *General:* Toe off (see Fig. 2.13) occurs at about 60 per cent of the gait cycle. It is the point at which the stance phase ends and the swing

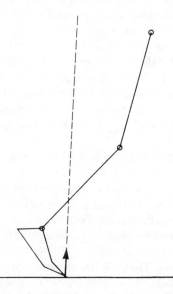

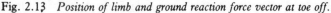

Fig. 2.13 *Position of limb and ground reaction force vector at toe off.*

phase begins. It also marks the end of the second double support phase of the cycle, since the other foot makes heel contact at about 50 per cent of the cycle. The period between heel off and toe off is sometimes called the *terminal rocker*, which is appropriate, since the leg is now rotating forwards about the forefoot rather than about the ankle joint. Another term for this period is the 'push off' phase. Perry (1974) objected to the use of this term because 'the late floor-reaction peak is the result of leverage by body alignment, rather than an active downward thrust.' However, Winter (1983) argued that the 'push off' not only exists, but is the period during which the generation of power by the muscles is greatest. The dispute can be resolved by comparing the gait of normal subjects and lower limb amputees, who are clearly incapable of producing an active 'push off'. Both the ground reaction force and the moment of force about the ankle joint are similar between these two groups. However, in normal subjects the ankle undergoes active plantarflexion during the time that the ankle moment is high, whereas in amputees there is only a small amount of passive plantarflexion, which occurs later when the load is removed from the foot. There thus seems little doubt that normal subjects do indeed have a 'push off'.

2. *Upper body:* The position of the upper body is roughly a mirror image of that described around foot flat, which is taking place on the other side. The extremes of arm and shoulder advancement, and the second low point for the trunk, all occur following heel off, and are beginning to reverse again by toe off.

3. *Hip:* The peak extension of the hip, which is between 10 and 20 degrees, is reached before toe off, by which time the hip is flexing again (see Fig. 2.5). Iliopsoas activity ceases before toe off, and the moment which flexes the hip is provided partly by gravity, partly by rectus femoris, in its dual role as hip flexor and knee extensor, and partly by the hip adductors, which also act as flexors when the hip is extended.

4. *Knee:* The knee starts to flex even before heel off, and by the time of toe off it is flexed to an angle of 40 to 50 degrees (see Fig. 2.5). Before the foot leaves the ground, the force vector moves from in front of the knee to behind it, and in the final part of the stance phase the knee is subjected to an external flexion moment (see Fig. 2.13). The knee is permitted to flex, but the rate is controlled by eccentric contraction of the rectus femoris, which flexes the hip at the same time that it controls knee flexion. The other elements of the quadriceps may also assist in controlling knee flexion. Knee flexion continues even after the foot has left the ground, partly as a result of the continued forward movement of the femur. During the swing phase the leg acts as a double pendulum, and knee flexion followed by extension takes place purely passively. Above-knee

amputees are able to walk with an almost normal swing phase, despite the absence of any muscles acting about the knee joint.

5. *Ankle and foot:* In the period between heel off and toe off, the ankle moves from dorsiflexion into plantarflexion. The total range of this movement varies between about 20 degrees (as in Fig. 2.5) and 35 degrees. Extension takes place at the metatarsophalangeal joints, as the heel and hindfoot lift up while the phalanges of the toes stay on the ground. The major force transmission to the ground is through the metatarsal heads. During this period there is a substantial external dorsiflexion moment about the ankle, which is resisted by powerful contraction of the soleus and gastrocnemius. The force transmitted up the leg flexes the knee and helps to initiate flexion of the hip. While load is being transmitted through the forefoot, the peronei, tibialis posterior and the long toe flexors contract to stabilize the foot and to permit the toes, as well as the metatarsal heads, to transmit force to the ground. As toe off approaches, the ground reaction force diminishes rapidly, and it disappears as the foot leaves the ground and the leg enters the swing phase. The foot is supinated at toe off, but this is largely lost during the first part of the swing phase.

Mid swing

1. *General:* Mid swing on one side (see Fig. 2.14) corresponds to mid stance on the other: it is the time when the swinging leg passes the stance phase leg, and the two feet are side by side. The swing phase occupies about 40 per cent of the gait cycle, and mid swing generally occurs close to the midpoint of this period. The swing phase is divided into two parts – an *acceleration phase*, before mid swing, and a *deceleration phase* following it. Alternative names are 'lift off' and 'reach' or 'initial swing' and 'terminal swing'. The walking velocity depends to a large extent on the efficiency of the swing phase, since the stride length is the amount by which the foot can be moved forwards during this time. If the foot catches on the ground, this will slow or even terminate the swing, and thereby reduce the stride length.

2. *Upper body:* At mid swing, as at mid stance, the trunk is at its highest position, and is maximally displaced over the stance phase leg. The arms are level with each other, one moving forward and one moving back.

3. *Hip:* The hip starts to flex even prior to toe off, and by the time of mid swing it has almost reached its most flexed position. Iliopsoas is active around heel off. It stops contracting briefly until toe off, and then contracts powerfully to flex the hip until mid swing. Hip flexion is aided

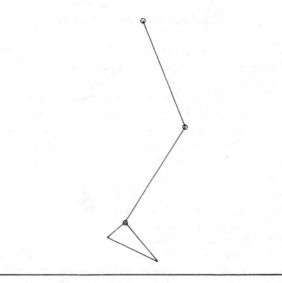

Fig. 2.14 *Position of limb at mid swing.*

by gravity, rectus femoris and the adductors. After mid swing, in the deceleration phase, the hamstrings contract, to slow down, stop and then reverse the flexion of the hip.

4. *Knee:* The flexion of the knee during the swing phase results largely from the flexion of the hip. The leg acts as a jointed pendulum, and no muscular contraction around the knee needs to be involved. The peak of swing phase flexion is usually between 60 and 70 degrees, and occurs before mid swing, by which time the knee has started to extend again. Muscular activity is required at the end of the swing phase, when the knee flexors, in particular the hamstrings, contract to prevent hyperextension of the knee, which would result from an uncontrolled continuation of the pendulum-like movement.

5. *Ankle and foot:* In normal walking, the toes clear the ground by very little. Murray (1967) found a mean clearance of 14 mm with a range of 1–38 mm. Most of the shortening of the leg required to achieve this comes from flexion of the knee, but the ankle also needs to move from its position of plantarflexion, at the end of the stance phase, to an approximately neutral position, to help with the clearance. This movement requires contraction of the anterior tibial muscles, although the force of contraction is much less than that required to control foot lowering following heel contact. The foot is supinated at toe off, and remains so until the following heel contact.

Muscular activity during gait

There is only partial agreement between investigators as to what is the 'normal' pattern of muscle usage during gait. Figure 2.15 shows a typical pattern, based largely on Perry (1974) and Inman et al. (1981). Similar, though not identical, data for these and other muscles were given by Winter (1987). A good review of the subject was provided by Shiavi (1985). The reason for the lack of general agreement is partly the inherent variability of muscle usage, and partly the limitations of the measuring technique. Such studies are normally made using EMG, a method of measurement which suffers from a number of problems, which will be further discussed in Chapter 4.

Although Fig. 2.15 shows a typical pattern, it is not the only possible one. One of the interesting things about gait is the way in which the same movement may be achieved in a number of different ways, and this particularly applies to the use of muscles, so that two people may walk with the same 'normal' gait pattern, but using different combinations of muscles. The pattern of muscle usage varies not only from one subject to another, but also with fatigue and with walking speed in a single person. The muscular system is said to possess 'redundancy', which means that if a particular muscle cannot be used, its functions may be taken over by another muscle or group of muscles.

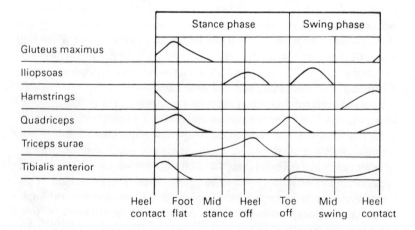

Fig. 2.15   *Typical activity of major muscle groups during the gait cycle.*

Optimization of energy usage

If we were fitted with wheels, very little energy would be needed for locomotion on a level surface, and some of the energy expended in going uphill would be recovered when coming down again. For this reason, both wheelchairs and bicycles are remarkably efficient forms of transport, although obviously much less versatile than a pair of legs. The legs have to be started and stopped, and the center of gravity of the body rises and falls and moves from side to side, all of which use energy. Despite this, walking is not as inefficient as it might be, due to two forms of optimization – those involving transfers of energy, and those which minimize the displacement of the center of gravity.

Energy transfers

Two types of energy transfer occur during walking – an exchange between potential and kinetic energy, and the transfer of energy between one limb segment and another. The most obvious exchange between potential and kinetic energy is in the movement of the trunk. During the double support phase, the trunk is at its lowest vertical position, with its highest forward velocity. During the first half of the single support phase, the trunk is lifted up by the supporting leg, converting some of its kinetic energy into potential energy as its velocity reduces. During the latter part of the single support phase, the trunk drops down again in front of the supporting leg, and reduces its height while picking up velocity again. These exchanges between potential and kinetic energy are the same as in a child's swing, in which the potential energy at the highest point in its travel is converted into kinetic energy as it swings downwards, then back into potential energy again as it swings up the other side.

As well as the vertical motion of the trunk, there are other exchanges between potential and kinetic energy in walking. The twisting of the shoulder girdle and pelvis in opposite directions stores potential energy, as tension in the elastic structures, which is converted to kinetic energy as the trunk untwists, and then back to potential energy again as the trunk twists the other way.

Winter et al. (1976) studied the energy levels of the limb segments and of the quaintly named HAT (Head, Arms and Trunk). These authors criticized earlier studies, which included the kinetic energy of linear motion but neglected the kinetic energy due to rotation, which is responsible for about 10 per cent of the total energy of the shank. They studied only the sagittal plane, regarding energy exchanges in the other planes as negligible. They confirmed the exchange between potential and kinetic energy, described above, and estimated that roughly half of the energy of the HAT segment was conserved in this way. The thigh conserved about a third of its energy by exchanges of this sort, and the shank virtually none. They also noted that the changes in total body energy were less than the changes in energy of the individual segments, indicating a transfer of energy from one segment to another. In one subject, during a single gait cycle, the energy changes were: shank 16 J, thigh 6 J and HAT 10 J, making a total of 32 J. However, the total body energy change was only 22 J, indicating a saving of 10 J by inter-segment transfers.

Power is the rate at which energy is generated or absorbed, and is measured in watts (W). Winter (1983) calculated the energy generated and absorbed across the different joints in walking. He found that the highest peak was during the push off phase, when the ankle plantarflexors generate around 500 W. Motion about the knee has three peaks of power absorption and one of power generation, all in the range 100–200 W. Power generation and absorption at the hip is more variable and generally below 100 W.

The six determinants of gait

The six optimizations used to minimize the excursions of the center of gravity were called the 'determinants of gait' in a classic paper by Saunders et al. (1953). A modified version was published by Inman et al. (1981). A brief description will be given, but one of the works cited should be consulted for a detailed and well illustrated account. The fourth and fifth determinants were combined in the original descriptions, but for the purposes of clarity they have been separated here. The six 'determinants of gait' are as follows:

1. *Pelvic rotation:* Starting with the leg vertical and the foot on the ground, any flexion or extension at the hip joint will not only move the trunk backwards or forwards, but will also lower its vertical height. The longer the stride length, the greater the angles of flexion and extension of the hip, and the more the trunk loses height. The first 'determinant of gait' is the way in which the pelvis twists about a vertical axis during the gait cycle, bringing the hip joint forwards as the hip flexes, and backwards as it extends. This means that for a given stride length, less flexion and extension of the hip is required, since a proportion of the stride length comes from the forward and backward movement of the hip joint rather than the angular movement of the leg. The reduction in the

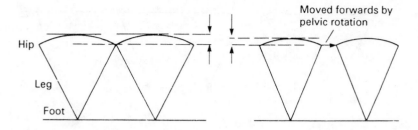

Fig. 2.16 *First determinant of gait: pelvic rotation reduces the angle of hip flexion and extension, which in turn reduces the vertical movement of the hip.*

range of hip flexion and extension leads to a reduction in the vertical movement of the trunk (Fig. 2.16).

2. *Pelvic tilt:* As described above, flexion and extension of the hip is accompanied by a rise and fall in the height of the hip joint. If the pelvis were to keep level, the trunk would follow this up and down movement. However, the second 'determinant of gait' is the way in which the pelvis tilts from side to side, so that when the hip of the stance phase leg is at its highest point, the pelvis is inclined, to lower the hip of the swing phase leg. The height of the trunk depends not on the height of one or other hip joint but on the average of the two of them, so the pelvic tilt reduces the total vertical excursion of the trunk (Fig. 2.17). This pelvic tilt can only be achieved if the swing phase leg can be shortened sufficiently (normally by flexing the knee and dorsiflexing the ankle) to clear the ground, despite a lowering in the height of its hip joint.

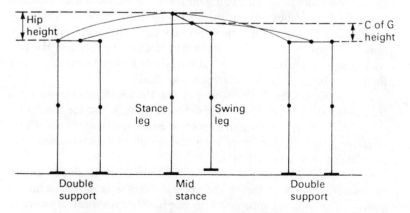

Fig. 2.17 *Second determinant of gait: the vertical movement of the center of gravity (C of G) is less than that of the hip, due to pelvic tilt.*

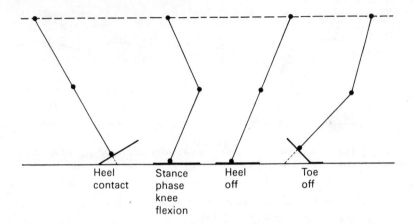

Heel   Stance   Heel   Toe
contact   phase   off   off
              knee
              flexion

Fig. 2.18   *Third, fourth and fifth determinants of gait: stance phase knee flexion shortens the leg (third); the heel lengthens it (fourth); so does the forefoot (fifth).*

3. *Knee flexion in stance phase:* The third, fourth and fifth determinants of gait (Fig. 2.18) are all concerned with adjusting the effective length of the leg during the stance phase, to keep the hip height as constant as possible. The third 'determinant' is the stance phase flexion of the knee. As the femur passes from flexion of the hip into extension, the hip joint would rise and then fall if the leg remained straight. However, flexion of the knee shortens the leg in the middle of this movement, reducing the height of the apex of the curve.

4. *Ankle mechanism:* Complementary to the way in which the apex of the curve is reduced by shortening the leg in the middle of the movement from hip flexion to extension, the beginning of the curve is elevated by lengthening the leg at the time of heel contact. This is achieved by the fourth 'determinant of gait' – the ankle mechanism. Because the heel sticks out behind the ankle joint, it effectively lengthens the leg during the period between heel contact and foot flat.

5. *Foot mechanism:* In the same way that the heel lengthens the leg at the start of the stance phase, the forefoot lengthens it at the end, in the fifth 'determinant' – the terminal rocker. From the time of heel off, the effective length of the lower leg increases as the ankle moves from dorsiflexion into plantarflexion.

6. *Lateral displacement of body:* The first five determinants of gait are all concerned with reducing the vertical excursions of the center of gravity. The sixth is concerned with side to side movement. By keeping the walking base narrow, little lateral movement is needed to preserve balance during walking (Fig. 2.19). The reduction in lateral acceleration

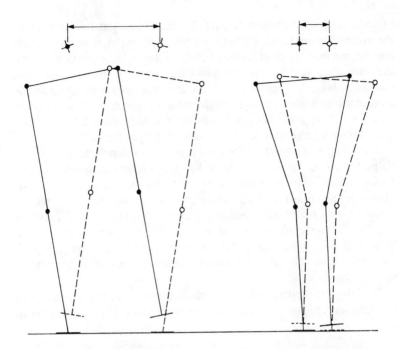

Fig. 2.19 *Sixth determinant of gait: having the feet closer together reduces the side-to-side movement of the center of gravity necessary to maintain balance.*

and deceleration leads to a reduction in the use of muscular energy. The main adaptation which allows the walking base to be narrow is a slight valgus angulation of the knee, which permits the tibia to be vertical while the femur inclines inwards, from a slightly adducted hip.

It is obvious that although the six determinants of gait have been described separately, they are all integrated together during each gait cycle. The combined effect is a much smoother trajectory for the center of gravity, and a much reduced energy expenditure, than would have been the case without them.

Ground reaction forces

The *force platform* (or forceplate) is an instrument commonly used in gait analysis. It gives the total force applied by the foot to the ground, although it does not show the distribution of this force across the sole of the foot. Some force platforms give only one component of the force (usually vertical), but most give a full three-dimensional description of the average ground reaction force vector. The electrical output signals may be processed to produce three components of force (vertical, lateral and fore–aft), the two coordinates of the center of pressure, and the moments about the vertical axis. The *center of pressure* is the point on the ground through which a single resultant force appears to act, although in reality the total force is made up of innumerable small force vectors, spread over a finite area on the surface of the platform.

Since the ground reaction force is a three-dimensional vector, it would be preferable to display it as such for the purposes of interpretation. Unfortunately, this is seldom practical. The most common form of display is shown in Fig. 2.20, where the three components of force are plotted against time for the walk which is shown in Figs 2.4 to 2.7 and 2.9 to 2.14. The sign convention used in Fig. 2.20 is the same as that used by Winter (1987), where the ground reaction force is positive upwards, forwards and to the right. Regrettably there is no general agreement on sign conventions.

The lateral component of force is small; for most of the stance phase of the right foot, the ground reaction force accelerates the center of gravity

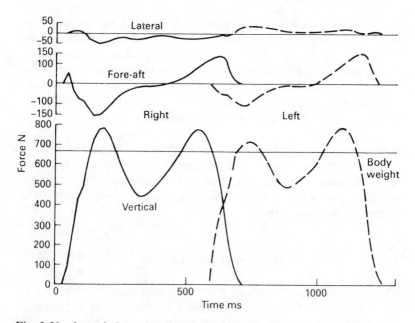

Fig. 2.20   *Lateral, fore–aft and vertical components of the ground reaction force, in newtons, for right foot (solid line) and left (dashed).*

towards the left side of the body, and during the stance phase of the left foot the acceleration is towards the right side of the body. The fore–aft (or antero–posterior) trace from the right foot shows the forward directed heelstrike transient, the 'braking' during the first half of the stance phase, and the 'propulsion' during the second half. The left foot shows the same pattern, but without the heelstrike transient. The force platform data were sampled at 20 ms intervals, and the heelstrike transient was probably missed because it occurred during a gap between successive data points. The vertical force shows a characteristic double hump, which results from an upward acceleration of the center of gravity during early stance, a reduction in downward force as the body 'flies' over the leg in mid stance, and a second peak due to deceleration as the downward motion is checked in late stance.

Plots of this type are difficult to interpret, and encourage consideration of the force vector as separate components rather than as a three-dimensional whole. The 'butterfly diagram' shown in Fig. 2.6 is an improvement on this, since it combines two of the force components (vertical and fore–aft) with the fore–aft coordinate of the center of pressure. It also preserves information on timing, since the lines representing the force vector are at regular intervals (20 ms in this case). Butterfly diagrams for the coronal and transverse planes are more difficult to interpret, and are seldom used.

The other type of information commonly derived from force platform data is the position of the center of pressure of the two feet on the ground, as shown in Fig. 2.21, again for the same walk. This may be used to

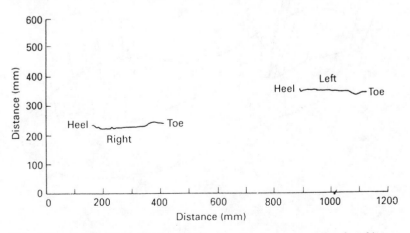

Fig. 2.21 *Center of pressure beneath the two feet in a normal female subject, walking from left to right.*

identify abnormal patterns of foot contact, including an abnormal toe out or toe in angle. The step length and walking base can also be measured from this type of display, providing there is a definite and identifiable heel contact.

If the pattern of foot contact is of particular interest, it is preferable to combine the data on the center of pressure with an outline of the foot obtained by some other means (such as chalk on the floor). This type of display, with the addition of a sagittal plane representation of the ground reaction force vector, is shown in Fig. 2.22, for a normal male subject wearing shoes. The trace shows initial contact at the back of the heel on the lateral side, with progression of the center of force along the middle of the foot to the metatarsal heads, where it moves medially, ending at the hallux. The spacing of the vectors shows how long the center of pressure spends in any one area. It is worth noting that there is a cluster of vectors just in front of the edge of the heel, where the shoe is not in contact with the ground, again pointing out the fact that the center of pressure is merely the average of a number of forces acting beneath the foot.

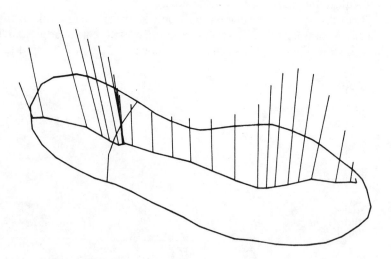

Fig. 2.22   *Foot outline, center of pressure and sagittal plane representation of ground reaction force vector; right foot of a normal male subject walking in shoes.*

The force platform has been described as a 'whole body accelerometer' – its output gives the acceleration in three-dimensional space of the center of gravity of the body as a whole, including both the limb which is on the ground and the leg which is swinging through the air.

Moments of force

Figure 2.23 shows the internal moments of force, in newtons per kilogram body weight, at the hip, knee and ankle during the gait cycle. The curves were derived from 19 subjects walking at normal cadence (Winter, 1987).

Ankle moment

The internal ankle moment is substantial and has a characteristic shape.

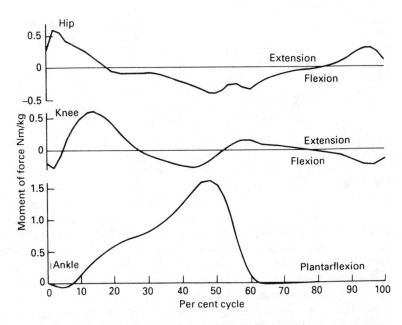

Fig. 2.23 *Internal moments about hip, knee and ankle joints, in newton-meters per kilogram body weight. Positive moments are generated by hip extensors, knee extensors and ankle plantarflexors. (Winter, 1987.)*

Between heel contact and foot flat, the foot is plantarflexed by a relatively small external moment, which is opposed by an internal dorsiflexion moment generated by the anterior tibial muscles. This is quickly replaced by a plantarflexion moment, which builds up as the triceps surae contracts, to a peak of about 110 N·m just before toe off. The moment plantarflexes the ankle, and propels the center of gravity of the body upwards and forwards. Once the foot has left the ground, a small dorsiflexion moment from the anterior tibial muscles is used to oppose the effects of gravity on the foot.

Knee moment

The internal knee moment in the stance phase usually commences with a brief flexion peak, prior to foot flat, as the hamstrings contract to prevent the knee from hyperextending on initial ground contact. This is followed by an extension moment, which peaks at around 40 N·m, produced by contraction of the quadriceps during the stance phase knee flexion. As the knee extends again, the moment reverses, a flexion moment of around 20 N·m being produced by contraction of the gastrocnemius. Just before toe off, the moment changes again, becoming extensor for the first part of the swing phase, mainly due to contraction of the rectus femoris component of the quadriceps. At the end of the swing phase, the knee extends rapidly due to the pendulum action of the leg, and hyperextension is prevented by another peak of flexion moment, from contraction of the hamstrings.

Hip moment

The internal hip moment shows that in the first third of the stance phase there is activity by the extensor muscles, particularly gluteus maximus, as the leg is pulled back into extension, in opposition to the external moment produced by the ground reaction force. As the stance phase progresses, the line of the ground reaction force moves further back, the external moment changes direction to become an extension moment, and an internal flexion moment is generated by eccentric contraction of the iliopsoas. As the hip begins to flex again just prior to toe off, the internal flexion moment continues but the eccentric contraction becomes concentric. At the peak of hip extension, the adductor muscles augment the iliopsoas as hip flexors. Towards the middle of the swing phase, the hip reaches its maximum flexion, and further flexion is prevented by an internal extension moment arising in the hamstrings.

Support moment

As well as the moments shown in Fig. 2.23, Winter (1980) described a 'support moment', which is the sum of the moments at the hip, knee and ankle. It is not a true algebraic sum, since the sign convention is based on flexion and extension, rather than on clockwise and counterclockwise moments, so that the direction of the knee moment is opposite to that of the hip and ankle moments.

Vector projection

A method commonly used to calculate joint moments is to project the ground reaction force vector upwards, to measure its distance from the joint center, and to multiply this distance by the magnitude of the force. As described previously, this method ignores both the inertial and the gravitational factors, and thus introduces errors. These errors are negligible as far as the ankle is concerned, and only amount to a few per cent for the knee. At the hip, however, the errors are substantial and it is unwise to use the method except as a very rough approximation. The reason this method is sometimes used is to avoid the full procedure for calculating joint moments, which is very complicated and requires detailed information on the mass, center of gravity and moment of inertia of each of the limb segments, as well as their linear and angular accelerations during the gait cycle.

Energy consumption

It is relatively easy to measure the energy consumption of a vehicle, but much more difficult to make equivalent measurements for human walking, for two reasons. Firstly, there is a clear relationship between the fuel level in the tank of a vehicle and how much energy has been used, whereas knowing how much food a person has eaten gives no information on the energy consumed in a particular activity. Secondly, a vehicle which is switched off uses no energy, whereas people use metabolic energy all the time, whether they are walking or not.

The first problem – that of measuring the 'fuel consumption' – can be solved by measuring not the fuel being consumed but the oxygen which is used to oxidize it. Measurements of oxygen uptake, while not particularly pleasant for the subject (who has to wear a face mask or mouthpiece) are nonetheless perfectly practical, and are used routinely to measure the metabolic cost of different activities.

The second problem – the lack of a suitable baseline for energy consumption measurements in humans – is not easy to solve and requires a different way of thinking about the subject. The energy used by a person who is walking can be divided into three parts:

1. The muscles used for walking consume energy, as they accelerate and decelerate the trunk and the limb segments in different directions.

2. There is an 'overhead' involved in walking, in that the expenditure of energy by the muscles involves increased activity by the heart and the lungs, which themselves use energy. Further energy is expended in maintaining the upright posture.

3. The 'basal metabolism' is the irreducible minimum energy a person will consume if totally at rest.

The relationship between metabolic energy and physical energy is very complicated. As explained in Chapter 1, if a muscle undergoes an isometric contraction, it still uses energy, although its length does not change and the physical work it does is zero. In an eccentric contraction, when a muscle lengthens under tension, it uses up metabolic energy, when in physical terms one would expect it to gain energy rather than to lose it.

In the past it has been usual to estimate the mechanical efficiency of walking by looking at the difference in oxygen consumption between the 'basal' state, and walking at a given speed. This approach neglects the various overheads, however, and makes slow walking appear to be extremely inefficient. Inman et al. (1981) suggested that it is more realistic to use standing or very slow walking as the baseline for measurements on faster walking. Despite these complications, a figure of 25 per cent is often quoted for the efficiency of the conversion of metabolic energy into mechanical energy, in a wide range of activities, including walking.

The energy requirements of walking can be expressed in two ways: the energy used per unit of time, or the energy used per unit of distance.

Energy consumption per unit time

Inman et al. (1981) quoted an equation, based on a number of studies, for the relationship between the walking speed and the energy consumption per unit time. The energy consumption included both the basal metabolism and the 'overheads'. They showed, not surprisingly, that energy consumption per unit time is less for slow walking than for fast walking. Translating their equation into SI units, it becomes:

$$E_w = 2.23 + 1.26\ v^2$$

where E_w is the energy consumption in watts per kilogram body mass, and v is the velocity in m/s.

As an example of the application of this equation, a 70 kg person walking at 1.4 m/s, which is a typical speed for adults, would consume energy at a rate of 330 W.

Energy consumption per unit distance

The energy consumption per meter walked has a less straightforward relationship with walking speed, as both very slow and very fast walking speeds use more energy than walking at intermediate speeds. The equation which describes this relationship, again converted into SI units, is:

$$E_m = 2.23/v + 1.26\,v$$

where E_m is the energy consumption in joules per meter per kilogram body mass, and v is the velocity in m/s.

The minimum energy usage is predicted by this equation at a velocity of 1.33 m/s. A 70 kg person walking at this speed would use 235 joules per meter, or 235 kJ per kilometer. A typical 'candy' bar contains around 1000 kJ, and would thus supply enough energy to walk 4.26 km or more than two and a half miles!

The equations quoted above merely give average values, which may be modified by age, sex, walking surface, footwear, and so on. Pathological gait is frequently associated with an energy consumption considerably above these 'average' values. To provide a baseline for studies of pathological gait, Waters et al. (1988) made a detailed study of the energy consumption of a total of 260 normal children and adults of both sexes, walking at a variety of speeds.

Gait in the young

Although a number of studies have been made of the development of gait in children, that by Sutherland et al. (1988) is the most detailed. The main ways in which the gait of small children differs from that of adults are as follows:

1. The walking base is wider
2. The stride length and velocity are lower and the cadence higher
3. Children have no heelstrike, initial contact being made by the flat foot
4. There is very little stance phase knee flexion

5. The whole leg is externally rotated during the swing phase
6. There is an absence of reciprocal arm swinging.

These differences in gait mature at different rates. The characteristics numbered (3), (4) and (5) in the above list have changed to the adult pattern by the age of two years, and (1) and (6) by the age of four. The cadence, stride length and velocity continue to change with growth, reaching normal adult values at around the age of 15 years.

Most children commence walking within three months of their first birthday. Prior to this, even small babies will make reciprocal stepping motions if they are moved slowly forwards while held in the standing position with their feet on the ground. However, this is not true walking, as there is little attempt to take any weight on the legs.

Figure 2.24, which is based on Sutherland et al. (1988), shows the average sagittal plane motion at the hip, knee and ankle joints in 49 children between 11 and 13 months. It should be compared with Fig. 2.5, which shows the same parameters for adult gait. Sutherland et al. do not give the timing of mid stance, heel off or mid swing.

The pattern of hip flexion and extension differs from that in adults in

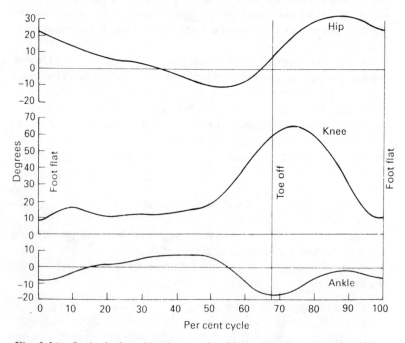

Fig. 2.24 *Sagittal plane hip, knee and ankle angles in one-year-old children. Sign conventions as in Fig. 2.5. (Sutherland et al., 1988.)*

that the degree of extension is reduced, and the hip does not remain flexed for so long at the end of the swing phase.

The knee never fully extends, but this is seen at all ages in Sutherland's data, and may reflect the method of measurement. There is some stance phase flexion in infants, but it is both smaller in magnitude and earlier than in adults. (It should be noted that most adults show more stance phase flexion than is seen in Fig. 2.5.) The flexion of the knee in the swing phase is also somewhat reduced at the age of one year.

Initial contact in small children is by the whole foot, heel contact being replaced by foot flat. The ankle is plantarflexed at initial contact, and remains so into the early stance phase, in contrast to the adult pattern, in which the ankle is approximately neutral at heel contact, but moves rapidly into plantarflexion. The pattern of dorsiflexion followed by plantarflexion through the remainder of the stance phase is essentially the same at all ages.

Since children are smaller than adults, it is not surprising that they walk with a shorter stride length and at a lower velocity. Sutherland et al. (1988) showed that stride length is closely related to height, and that the ratio of stride length to stature is similar to that found in adults. The change in stride length with age mirrors the change in height, showing a rapid increase up to the age of four years, and a slower increase thereafter. Todd et al. (1989) showed the relationships between the height of children and the general gait parameters. Small children walk with a rapid cadence, the mean at the age of one being about 170 steps per minute. It reduces with age, but is still around 140 steps per minute at age seven, which is well above the average adult values of 113 for men and 118 for women. The higher cadence partly compensates for the short stride length, and the velocity ranged from 0.64 m/s at age one to 1.14 m/s at age seven, compared with the typical adult values of 1.46 m/s for males and 1.30 m/s for females. Sutherland did not report on the gait of children beyond the age of seven, and did not distinguish between the results from boys and girls. Appendix 1 gives the normal ranges for the general gait parameters in children, derived in part from Sutherland's data.

As can be seen in Fig. 2.24, the swing phase occupies a smaller proportion of the gait cycle in very small children than in adults, which minimizes the time spent in the less stable condition of single legged stance. The relative duration of the swing phase increases with age, reaching the adult proportion around the age of four years. There is symmetry between the two sides at all ages. Sutherland et al. (1988) related the width of the walking base to the width of the body at the top of the pelvis, giving a slightly confusing 'pelvic-span:ankle-spread' ratio.

Changing the measurement units for the sake of clarity, the walking base is about 70 per cent of the pelvic width at the age of one year, falling to about 45 per cent by the age of three and a half, at which level it remains until the age of seven. An average value for adults is not readily available, but it is probably less than 30 per cent.

At the very youngest ages, the EMG patterns showed that there is a tendency to activate most muscles for a higher proportion of the gait cycle than in adults. With the exception of the triceps surae, adult patterns are established for most muscles by the age of two years. Sutherland et al. (1988) found that children could be divided into two groups depending on whether the triceps surae was activated in a prolonged (infant) pattern or the normal (adult) pattern. Below the age of two years, over 60 per cent of the children showed the infant pattern; the proportion dropped to below 30 per cent by the age of seven. They speculated that this might relate to delayed myelination of the sensory branches of the peripheral nerves.

Gait in the elderly

A number of investigations have been made of the changes in gait which occur with advancing age, especially by Murray et al. (1969), who studied the gait of men up to the age of 87. The description which follows is confined to the effects of age on free-speed walking, although Murray et al. also examined fast walking. A companion paper (Murray et al., 1970) studied the gait of women up to age 70. It did not provide as much information on the effects of age, but generally confirmed the observations made on males.

The gait of the elderly is subject to two influences – the effects of age itself, and the effects of pathological conditions such as osteoarthritis and parkinsonism, which become more common with advancing age. Providing patients with such conditions are carefully excluded, the gait of the elderly appears to be simply a 'slowed down' version of the gait of younger adults. Murray et al. (1969) were careful to point out that 'the walking performance of older men did not resemble a pathological gait.'

Typically, the onset of age-related changes in gait takes place at 60 to 70 years of age. There is a decreased stride length, a variable but generally decreased cadence, and an increase in the walking base. Many other changes can also be observed, but most of them are secondary to these three. The velocity, being the product of stride length and cadence, is almost always reduced in elderly people. Appendix 1 gives normal ranges for the general gait parameters up to the age of 80 years.

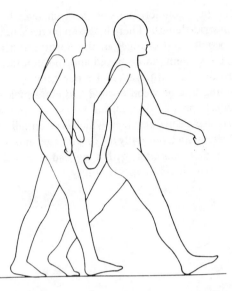

Fig. 2.25 *Body position at right heel contact in older men (left) and younger men (right) (Murray et al., 1969.)*

Some of the differences between the gait of the young and the elderly are apparent in Fig. 2.25, which is taken from Murray et al. (1969). These authors suggested that the purpose of the gait changes in the elderly is to improve the security of walking. Both decreasing the stride length and increasing the walking base make it easier to maintain balance while walking. Reducing the cadence leads to a reduction in the percentage of the gait cycle for which there is only single limb support, since the increase in cycle length is largely achieved by lengthening the stance phase and hence the double-support time.

Changes in the angular excursions of the joints in the elderly include a reduction in the total range of hip flexion and extension, a reduction in swing phase knee flexion, and reduced ankle plantarflexion during the push off. However, all of these depend on both cadence and stride length, and are probably within normal limits if these factors are taken into account. The vertical movement of the head is reduced and its lateral movement increased, probably secondary to the changes in stride length and walking base, respectively.

The trajectory of the toe over the ground is modified in old age, giving an improved ground clearance during the first half of the swing phase. This is probably another mechanism for improving security. The heel rises less during the push off phase, and the foot attitude is closer to the horizontal at heel contact, both of these changes being related to the reduction in stride length. There is also an increase in the angle of toe out in elderly people, and changes in the posture and movements of the arms, the elbows being more flexed and the shoulders more extended. The reasons for these differences are not known.

The dividing line between normal and abnormal may be difficult to define in elderly people. A condition known as 'idiopathic gait disorder of the elderly' has been described, which is essentially an exaggeration of the gait changes which normally occur with age, and is characterized by a cautious attitude to walking, with a low cadence, a short stride length, and an increased step-to-step variability.

3

Pathological Gait

Although some variability is present in normal gait, particularly in the use of the muscles, there is a clearly identifiable 'normal pattern' of walking, and a 'normal range' can be defined for most of the measurable parameters. Pathology of the locomotor system frequently produces abnormal patterns of movement or applied force. Some of these abnormalities can be identified by eye, but others can only be identified by the use of appropriate measurement systems.

In order that a person can walk, the locomotor system must be able to accomplish four things:

1. Each leg must be able to support the body weight without collapsing
2. Balance must be maintained, either statically or dynamically, during single leg stance
3. The swinging leg must be able to advance to a position where it can take over the supporting role
4. Sufficient power must be provided to make the necessary limb movements and to advance the trunk.

In normal walking, all of these are achieved without any apparent difficulty, and with a modest energy consumption. However, in many forms of pathological gait they can be accomplished only by means of abnormal movements, which usually increase the energy consumption, or by the use of walking aids such as canes, crutches and orthotic devices (callipers and braces). Failure to achieve all four requirements means that the subject is unable to walk.

The pattern of gait is the outcome of a complex interaction between the many neuromuscular and structural elements of the locomotor system. Abnormal gait may result from a disorder in any part of this system, including the brain, spinal cord, nerves, muscles, joints and skeleton. Abnormal gait may also result from the presence of pain, so that although physically capable of walking normally, the subject finds it more comfortable to walk in some other way.

The term *limp* is commonly used to describe a wide variety of abnormal gait patterns. However, dictionary definitions are unhelpful, a typical one being 'to walk lamely'. Since the word has no defined scientific meaning, it should only be used with caution in the context of gait analysis. The most appropriate use of the word is for a gait abnormality involving some degree of asymmetry, which is readily detected by the untrained observer.

Since gait is the end result of a complicated process, a number of different original problems will manifest themselves in the same abnormality of gait. For this reason, the abnormal gait patterns are described separately from the pathological conditions which cause them. The first part of the chapter describes, in some detail, the commonest abnormal gait patterns. This is followed by a description of the use of walking aids, and the gait of amputees. The chapter ends with a description of the way in which gait abnormalities may result from a number of common but serious neurological conditions.

Specific gait abnormalities

The following section is based on a manual of lecture notes for student orthotists produced by New York University (1986). The manual includes a very useful list of common gait abnormalities, all of which can be identified by eye. The manual criticizes the common practice of identifying gait abnormalities by their pathological cause, for example 'hemiplegic gait.' This immediately suggests that all hemiplegics walk in the same way, which is far from true, and also neglects the changes in gait which may occur with time or treatment. The manual suggests that it is preferable to use purely descriptive terms, such as 'excessive medial foot contact.' This practice will be adopted in the following descriptions. Some of the gait abnormalities described in the New York University publication applied to the gait of subjects wearing orthotic devices; these descriptions have been omitted from the present text.

The pathological gait patterns to be described may occur either alone or in combination. If in combination, they may interact, so that the individual gait modifications do not exactly fit the description. The list which follows is not exhaustive – a subject may use a variation of one of the general patterns, or may use some other gait pattern, which is not listed here.

When studying a pathological gait, particularly one which does not appear to fit into one of the standard patterns, it may be helpful to remember that an abnormal movement may be performed for one of two reasons:

1. The subject has no choice, the movement being 'forced' on them by weakness, spasticity or deformity.

2. The movement is a compensation, which the subject is using to correct for some other problem, which therefore needs to be identified.

Lateral trunk bending

Bending the trunk towards the side of the supporting limb during the stance phase is known as lateral trunk bending, or more commonly a *Trendelenburg gait*. The purpose of the maneuver is to reduce the forces in the abductor muscles and hip joint during single leg stance. Some authorities distinguish between the true Trendelenburg gait, in which there is hip abductor weakness, and the 'antalgic' gait of someone with a painful hip. However the term 'Trendelburg gait' is commonly applied to both circumstances.

Lateral trunk bending is best observed from the front or the back. During the double support phase, the trunk is generally upright, but as soon as the swing leg leaves the ground, the trunk leans across towards the side of the stance phase leg, returning to the upright attitude again at the beginning of the next double support phase. The trunk bending is frequently unilateral, being restricted to the stance phase of one leg, although it may be bilateral, the trunk swaying from side to side, to produce a gait pattern known as *waddling*.

Figure 3.1 shows a schema of the trunk, pelvis and hip joints when

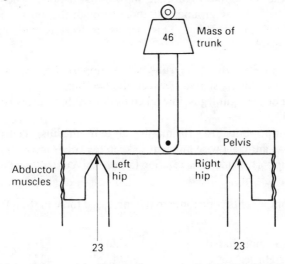

Fig. 3.1   *Schema of double legged stance: the force in each hip joint is half the mass of the trunk. Note: To make the diagrams easier to understand, masses are given in kg and forces in kgf (1 kgf = 9.81 N).*

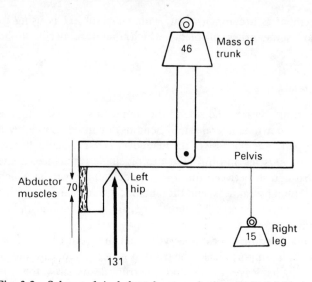

Fig. 3.2 *Schema of single legged stance: the force in the left hip is the sum of the mass of the trunk, the mass of the right leg, and the contraction force of the abductor muscles (see note to Fig. 3.1).*

standing on both legs, with typical values for the mass of the trunk and the forces on the hip joints. Figure 3.2 shows what happens when the right foot is lifted off the ground: the force through the left hip joint increases from 23 kgf to 131 kgf. This increase in force is made up of three components:

1. The whole of the weight of the trunk is now supported by the left hip joint, instead of being shared between the two hips.
2. The weight of the right leg is now taken by the left hip, instead of by the ground.
3. The left hip abductors (primarily gluteus medius) contract, producing an anticlockwise moment to keep the pelvis from dipping on the unsupported side. The reaction force to this contraction passes through the left hip.

These three components contribute to the increased force in the left hip as follows:

1. Trunk weight not shared:	226 N	(23 kgf)
2. Weight of right leg:	147 N	(15 kgf)
3. Contraction of abductors:	687 N	(70 kgf)
Total:	1060 N	(108 kgf)

It should be noted that this example was invented, for the purposes of illustration, so the actual numbers should not be taken too seriously!

The four conditions which must be met if this mechanism is to operate satisfactorily are:

1. The absence of significant pain on loading
2. Adequate power in the hip abductors
3. A sufficiently long lever arm for the hip abductors
4. A solid and stable fulcrum in or around the hip joint.

Should one or more of these conditions fail to be met, the subject may adopt lateral trunk bending in an attempt to compensate.

The effect on the joint force of lateral trunk bending is shown in Fig. 3.3. There is no effect on components (1) and (2) of the increased force, but if the center of gravity of the trunk is moved directly above the left hip, this eliminates the clockwise moment produced by the mass of the trunk. The abductors are now only required to contract with a force of 30 kgf, to balance the clockwise moment provided by the weight of the right leg. There is thus a reduction of 40 kgf in the abductor contraction

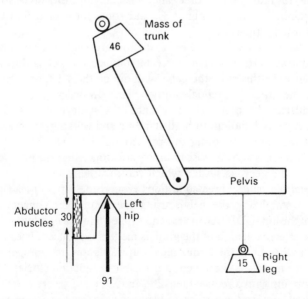

Fig. 3.3 *Lateral trunk bending reduces the clockwise moments about the left hip, permitting the pelvis to be stabilized by a smaller abductor force, with a corresponding reduction in hip joint force (see note to Fig. 3.1).*

force, and a corresponding reduction in the total joint force. The numbers in the illustrations refer to standing. During the stance phase of walking, higher forces are to be expected, due to the vertical accelerations of the center of gravity, which cause the force transmitted through the leg to fluctuate above and below body weight (see Fig. 2.20). However, this effect may be less in pathological than in normal gait, since these vertical accelerations are less in someone walking with a slower cadence and a shorter stride length. The numbers also suppose that the bending of the trunk brings its center of gravity exactly above the hip joint. This is unlikely to happen in practice, of course, but the principles remain the same, whether the center of gravity fails to reach the hip joint or even passes lateral to it.

There are a number of conditions in which this gait abnormality is adopted:

1. *Painful hip:* If the hip joint is painful, as in osteoarthritis and rheumatoid arthritis, the amount of pain experienced usually depends to a very large extent on the force being transmitted through the joint. Since lateral trunk bending reduces the total joint force, 'Trendelenburg gait' is extremely common in people with arthritis of the hip. Although it produces a useful reduction in force, and hence in pain, the forces still remain substantial (91 kgf in Fig. 3.3), and some form of definitive treatment is usually required.

2. *Hip abductor weakness:* If the hip abductors are weak, they may be unable to contract with sufficient force to stabilize the pelvis during single leg stance. In this case, the pelvis will dip on the side of the foot which is off the ground (Trendelenburg's sign). In order to give the weakened muscles the best possible chance, the subject will usually employ lateral trunk bending, in both standing and walking, to reduce the moment they are attempting to oppose (Fig. 3.4). Hip abductor weakness may be caused by disease or injury affecting either the muscles themselves, or the nervous system which controls them.

3. *Abnormal hip joint:* Three conditions around the hip will lead to difficulties in stabilizing the pelvis using the abductors: congenital dislocation of the hip (CDH), coxa vara and slipped femoral epiphysis. In all three, the effective length of the gluteus medius is reduced, because the greater trochanter of the femur moves upwards towards the pelvic brim. Since the muscle is shortened, it is unable to function efficiently, and thus contracts with a reduced tension. In CDH and severe cases of slipped femoral epiphysis, a further problem exists in that the normal hip joint is effectively lost, to be replaced by a false hip joint, or 'pseudarthrosis'. This abnormal joint is more laterally placed, giving a

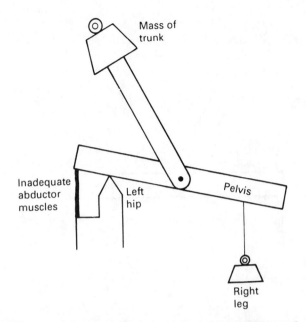

Fig. 3.4 *Trendelenburg's sign: due to inadequate hip abductors, the pelvis drops on that side when one foot is lifted off the ground. To compensate, the subject bends the trunk over the supporting hip.*

reduced lever arm for the abductor muscles, and it may fail to provide the 'solid and stable fulcrum' mentioned above. The combination of the reduced lever arm and the reduced power of muscular contraction gives these subjects a powerful incentive to walk with lateral trunk bending (Fig. 3.5). In many cases, particularly in older people with CDH, the false hip joint becomes arthritic, and they add a painful hip to their other problems. Pain is frequently also a factor in slipped femoral epiphysis.

4. *Wide walking base:* If the walking base is abnormally wide, there is a problem with balance during single leg stance. Rather than tip the whole body to maintain balance, as in the left-hand diagram in Fig. 2.19, lateral bending of the trunk may be used to keep the center of gravity of the body roughly over the supporting leg. In most cases, this will need to be done during the stance phase on both sides, leading to bilateral trunk bending and a *waddling* gait. A wide walking base is a gait abnormality for which a number of causes exist, which will be described later.

5. *Unequal leg length:* When walking with an unequal leg length, the pelvis tips downwards on the side of the shortened limb, as the body weight is transferred to it. This pelvic tilt is accompanied by a compensatory lateral bend of the trunk.

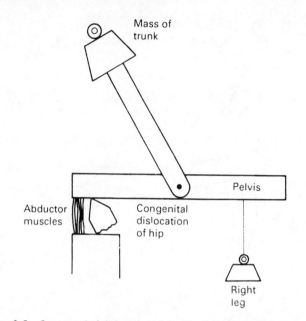

Fig. 3.5   *In congenital dislocation of the hip, both the working length and the lever arm of the hip abductors are reduced. To compensate, the subject bends the trunk over the supporting hip.*

Anterior trunk bending

In anterior trunk bending, the subject flexes his trunk forwards at the time of heel contact. If only one leg is affected, the trunk is straightened again around the time of heel contact by the other foot, but if both sides are affected, the trunk may be kept flexed throughout the gait cycle. The gait abnormality is best seen from the side.

The purpose of this gait pattern is to compensate for an inadequacy of the knee extensors. The left-hand diagram in Fig. 3.6 shows that between heel contact and foot flat, the line of action of the ground reaction force vector normally passes behind the axis of the knee joint and generates an external moment which attempts to flex it. This is opposed by contraction of the quadriceps, to generate an internal extension moment. If the quadriceps is weak or paralyzed, it cannot generate this internal moment, and the knee will tend to collapse. As shown in the right-hand diagram of Fig. 3.6, anterior trunk bending is used to move the center of gravity of the body forwards, which results in the line of force passing in front of the axis of the knee, producing an external extension (or hyperextension) moment. In addition to anterior

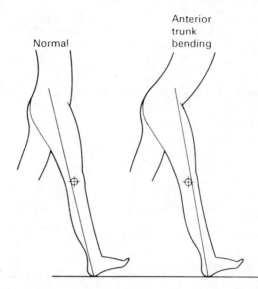

Fig. 3.6   *In normal walking, the line of force in early stance phase passes behind the knee. Anterior trunk bending brings the line of force in front of the knee, to compensate for weak knee extensors.*

trunk bending, subjects will often keep one hand on the affected thigh while walking, to provide further stabilization for the knee.

Posterior trunk bending

Posterior trunk bending is essentially a reversed version of anterior trunk bending, in that around the time of heel contact the whole trunk moves in the sagittal plane, but this time backwards instead of forwards. Again, it is most easily observed from the side.

The purpose of posterior trunk bending is to compensate for ineffective hip extensors early in the stance phase. The line of the ground reaction force at this time passes in front of the hip joint (Fig. 3.7 left). This produces an external moment which attempts to flex the trunk forward on the thigh, and is normally opposed by contraction of the hip extensors, particularly gluteus maximus. Should these muscles be weak or paralyzed, the subject compensates by moving the trunk backwards at this time, bringing the line of action of the external force behind the axis of the hip joint, as shown in the right-hand diagram of Fig. 3.7.

A different type of posterior trunk bending may occur early in the swing phase, where the subject may throw the trunk backwards in order

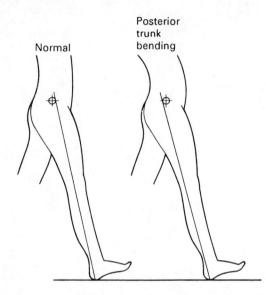

Fig. 3.7 *In normal walking, the line of force in early stance phase passes in front of the hip. Posterior trunk bending brings the line of force behind the hip, to compensate for weak hip extensors.*

to propel the swinging leg forwards. This is most often used to compensate for weakness of the hip flexors, or spasticity of the hip extensors, either of which makes it difficult to accelerate the femur forwards at the beginning of swing. This maneuver may also be used if the knee is unable to flex, since the whole leg must be accelerated forwards as one unit, which greatly increases the demands on the hip flexors. Posterior trunk bending may also occur when the hip is ankylosed, the trunk moving backwards as the thigh moves forwards.

Increased lumbar lordosis

Many people have an exaggerated lumbar lordosis, but it is only regarded as a gait abnormality if the lordosis is used to aid walking in some way, which generally means that the degree of lordosis varies over the course of the gait cycle. Increased lumbar lordosis is observed from the side of the subject, and generally reaches a peak at the end of the stance phase on the affected side.

The most common cause of increased lumbar lordosis is a flexion contracture of the hip. It is also seen if the hip joint is immobile due to ankylosis. This deformity causes the stride length to be very short by

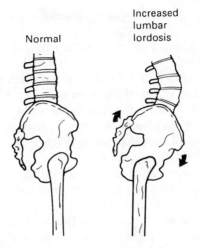

Normal

Increased
lumbar
lordosis

Fig. 3.8 *When there is a fixed flexion deformity of the hip, the whole pelvis must rotate forwards, with an increase in lumbar lordosis, for the femur to move into the extended position.*

preventing the femur from moving backwards from its flexed position. This difficulty can be overcome if the femur can be brought into the vertical (or even extended) position, not through movement at the hip joint, but by extension of the lumbar spine with a consequent increase in the lumbar lordosis (Fig. 3.8).

The orientation of the pelvis in the sagittal plane is maintained by the opposing pulls of the trunk muscles above and the limb muscles below. If there is a weakness of the muscles of the anterior abdominal wall, or of the hip extensors, or both, the pelvis is likely to tip anteriorly, again with an increase in the lumbar lordosis.

Functional leg length discrepancy

Four gait abnormalities (circumduction, hip hiking, steppage and vaulting) are closely related, in that they are designed to overcome the same problem – a functional discrepancy in leg length.

A 'functional' leg length discrepancy means that the legs are not necessarily different lengths when measured on the examination table, but that one or both are unable to adjust to the appropriate length for a particular phase of the gait cycle. This gait abnormality is frequently the result of a neurological problem. In order for natural walking to occur, the stance phase leg needs to be longer than the swing phase leg. If it is not, the swinging leg collides with the ground, and is unable to pass the stance leg.

The way that a leg is functionally lengthened is to extend at the hip and knee, and to plantarflex at the ankle. Conversely, the way in which a leg is functionally shortened is to flex at the hip and knee, and to dorsiflex at the ankle. Spasticity of any of the extensors or weakness of any of the flexors tends to make a leg too long in the swing phase, as does the mechanical locking of a joint in extension. Conversely, spasticity of the flexors or weakness of the extensors, or a flexion contracture in a joint, makes the limb too short for the stance phase.

An increase in functional leg length is particularly common following a 'stroke', where a foot drop may be accompanied by an increase in tone in the hip and knee extensor muscles.

The gait modifications designed to overcome the problem may either lengthen the stance phase leg or shorten the swinging leg, thus allowing a normal swing to occur. They are not mutually exclusive, and a subject may use them in combination. The gait modification employed by a particular person may have been forced on them by the underlying pathology, or it may have been a matter of chance, two people with apparently identical clinical conditions having found different solutions to the problem.

Circumduction

Ground contact by the swinging leg can be avoided if it is swung outward in a movement known as circumduction (Fig. 3.9). The swing phase of the other leg will usually be normal. The movement of circumduction is best seen from in front or behind.

Circumduction may also be used to advance the swinging leg in the presence of weak hip flexors, as the abductor muscles will also act as flexors while the hip joint is in the extended position.

Hip hiking

Hip hiking is a rather ugly gait modification, in which the pelvis is lifted on the side of the swinging leg (Fig. 3.10), by contraction of the spinal muscles and the lateral abdominal wall. The movement is best seen from behind or in front.

By tipping the pelvis up on the side of the swinging leg, hip hiking involves a reversal of the second determinant of gait. It may also involve an exaggeration of the first determinant (pelvic twist about a vertical axis)

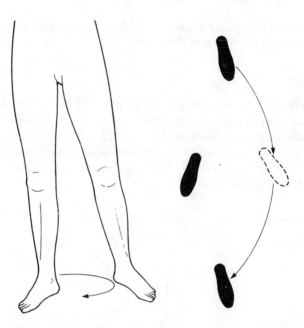

Fig. 3.9 *Circumduction: the swinging leg moves in an arc, rather than straight forwards, to increase the ground clearance.*

to assist with leg advancement. Leg advancement may also be helped by posterior trunk bending at the beginning of the swing phase.

·According to the manual by New York University (1986), hip hiking is commonly used in slow walking with weak hamstrings, since the knee tends to extend prematurely, and thus to make the leg too long towards

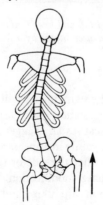

Fig. 3.10 *Hip hiking: the swing phase leg is lifted by raising the pelvis on that side.*

the end of the swing phase. It is seldom employed for limb lengthening due to plantarflexion of the ankle.

Steppage

Steppage is a very simple swing phase modification, consisting of exaggerated knee and hip flexion to lift the foot higher than usual for increased ground clearance (Fig. 3.11). It is best observed from the side. It is particularly used to compensate for a plantarflexed ankle, known colloquially as 'foot drop', due to inadequate dorsiflexion control which will be described later.

Fig. 3.11   *Steppage: increased hip and knee flexion improve ground clearance for the swing phase leg, in this case necessitated by a foot drop.*

Vaulting

The ground clearance for the swinging leg will be increased if the subject goes up on the toes of the stance phase leg, a movement known as vaulting (Fig. 3.12). This causes an exaggerated vertical movement of the trunk, which is both ungainly in appearance and wasteful of energy. It may be observed from either the side or the front.

Vaulting is a stance phase modification, whereas the related gait abnormalities (circumduction, hip hiking and steppage) are swing phase modifications. For this reason, vaulting may be a more appropriate solution for problems involving the swing phase leg. Like hip hiking, it is

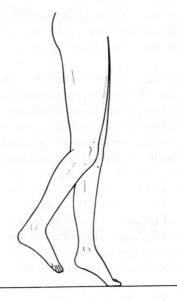

Fig. 3.12 *Vaulting: the subject goes up on the toes of the stance phase leg to improve ground clearance for the swinging leg.*

commonly used in slow walking with hamstring weakness, when the knee tends to extend too early in the swing phase.

Abnormal hip rotation

Because the hip is able to make large rotations in the transverse plane, for which the knee and ankle cannot compensate, an abnormal rotation at the hip involves the whole leg, with the foot showing a 'toe in' or 'toe out'. The gait pattern may involve both stance and swing phases, and is best observed from behind or in front.

Abnormal hip rotation may result from one of three causes:

1. A problem with the muscles producing hip rotation
2. A fault in the way the foot makes contact with the ground
3. As a compensatory movement to overcome some other problem.

Problems with the muscles producing hip rotation usually involve spasticity or weakness of the muscles which rotate the femur about the hip joint. For example, overactivity of the leg extensors in cerebral palsy may include an element of internal rotation. Weakness of biceps femoris will permit the medial hamstrings to rotate the femur medially, and conversely, weakness of the medial hamstrings will result in a lateral rotation of the leg. In cerebral palsy, the overactivity of extensor muscles, known as 'extensor synergy' frequently includes an element of internal rotation of the hip.

A number of foot disorders will produce an abnormal rotation at the hip. Inversion of the foot, whether due to a fixed inversion (pes varus) or to weakness of the peroneal muscles, will internally rotate the whole limb when weight is taken on it. A corresponding eversion of the foot, whether fixed (pes valgus) or due to weakness of the anterior and posterior tibial muscles, will result in an external rotation of the leg.

External rotation may be used as a compensation for quadriceps weakness, to alter the direction of the line of force through the knee. This could be used as an alternative to, or in addition to, anterior trunk deviation. External rotation may also be used to facilitate hip flexion, using the adductors as flexors, if the true hip flexors are weak. Subjects with weakness of the triceps surae may externally rotate the leg, to permit the use of the peroneal muscles as plantar flexors.

Excessive knee extension

In the gait abnormality of excessive knee extension, the normal stance phase flexion of the knee is lost, to be replaced by full extension or even hyperextension, in which the knee is angulated backwards. This is best seen from the side.

One cause of knee hyperextension has already been described: quadriceps weakness, which can be compensated for by keeping the leg fully extended, using anterior trunk deviation (Fig. 3.6) and/or external rotation of the leg to keep the line of the ground reaction force from passing behind the axis of the knee joint. Other means of keeping the knee fully extended are: pushing the thigh back by keeping one hand on it while walking; and using the hip extensors to snap the thigh sharply back at the time of heel contact.

Hyperextension of the knee, accompanied by anterior trunk bending, is seen quite frequently in people with paralysis of the quadriceps following poliomyelitis. The gait abnormality is clearly of great value to the subject, since without it he or she would be unable to walk. However, the hyperextension moment is resisted by tension in the posterior joint capsule, which gradually stretches, allowing the knee to develop a hyperextension deformity ('genu recurvatum'). As a result of this deformity, the joint frequently develops osteoarthritis in later life.

This is illustrated in Fig. 3.13, which shows the left hip, knee and ankle angles during walking in a 41-year-old lady whose quadriceps were

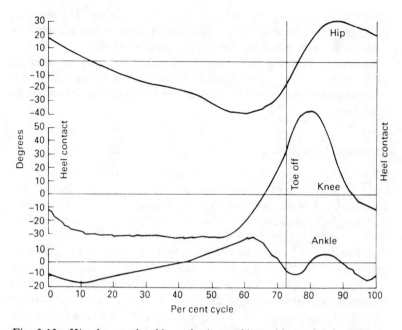

Fig. 3.13 *Hip, knee and ankle angles in a subject with paralyzed quadriceps, showing gross hyperextension of the knee and increased extension of the hip (cf Fig. 2.5).*

paralyzed on both sides by poliomyelitis at the age of 12. She walked very slowly, using two forearm crutches (cadence 63 steps/min; stride length 1.00 m; velocity 0.52 m/s). The knee hyperextends to 32 degrees during weightbearing, but flexes normally to 63 degrees during the swing phase. The hip flexes more than in normal individuals, since the hyperextension of the knee places it further back relative to the hip joint.

In normal walking, an external moment attempts to hyperextend the knee towards the end of the stance phase. This is resisted by an internal moment, generated by the knee flexors. Should the knee flexors be weak, the knee may be pushed backwards into hyperextension. If the triceps surae are weak or paralyzed, this may help with walking, since the leg lengthening required during the push off phase can be provided by extending the knee rather than by plantarflexing the ankle. However, there is a risk that the knee will go beyond full extension into hyperextension.

Hyperextension of the knee is common in spasticity, due to overactivity of the quadriceps. This may be accompanied by spasticity of the triceps surae, which plantarflexes the ankle and causes the body weight to be taken on the toes, even in normal standing. The resulting forward displacement of the ground reaction force vector generates an external moment which further hyperextends the knee.

Shortness of one leg may cause the person to stand on the other leg alone, with the knee hyperextended. This is because it is uncomfortable to stand on both legs, since the knee on the longer side would have to be kept flexed.

Excessive knee flexion

The knee is normally fully extended (or nearly so) twice during the gait cycle – around heel contact and around mid stance. In the gait abnormality known as excessive knee flexion, one or both of these movements into extension fails to occur. The flexion and extension of the knee are best seen from the side of the subject.

A flexion contracture of the knee will obviously prevent it from extending normally. A flexion contracture of the hip may also prevent the knee from extending, if hip flexion prevents the femur from becoming vertical or extended during the latter part of the stance phase (Fig. 3.14). By reducing the effective length of the leg during the stance phase, one of the compensations for a functional discrepancy in leg length may also be required.

Spasticity of the knee flexors may also cause the gait pattern of

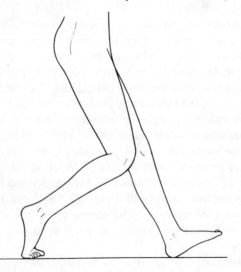

Fig. 3.14 *Increased knee flexion in late stance phase, necessitated by a flexion contracture of the hip.*

excessive knee flexion. Since the knee flexors are able to overpower the quadriceps, this may lead to other gait modifications, such as anterior trunk bending, to compensate for a relative weakness of the quadriceps.

The knee may flex excessively following heel contact if the normal plantarflexion of the foot between heel contact and foot flat is prevented through immobility of the ankle joint or a calcaneus deformity of the foot.

Increased flexion of the knee may be part of a compensatory movement, either to reduce the effective limb length in functional leg length discrepancy, or as part of a pattern of exaggerated hip, knee and arm movements, to make up for a lack of plantarflexor power in push off.

Inadequate dorsiflexion control

The dorsiflexors are active at two different times during the gait cycle, so that inadequate dorsiflexion control gives rise to two distinct gait abnormalities. Between heel contact and foot flat, the dorsiflexors resist the external plantarflexion moment, thus permitting the foot to be lowered gently. If they are weak, the foot is lowered abruptly in a *foot slap*. The dorsiflexors are also active during the swing phase, when they are used to raise the foot and achieve ground clearance. Failure to raise the foot sufficiently during the early swing phase may cause *toe drag*. Both problems are best observed from the side of the subject, and both make a distinctive noise. A subject with inadequate dorsiflexion control can often be diagnosed by ear, before they have come into view!

Inadequate dorsiflexion control may result from weakness or paralysis of the anterior tibial muscles, or from the dorsiflexors being overpowered by spasticity of the triceps surae. An inability to dorsiflex the foot during the swing phase is a functional leg length discrepancy for which a number of compensations have been described previously. Toe drag will only be observed if the subject fails to compensate. Toe drag may also occur if there is delayed flexion of the hip or knee at the beginning of swing, causing the foot to catch on the ground despite adequate dorsiflexion at the ankle.

Even if they suffer from inadequate dorsiflexion control, subjects with spasticity are frequently able to achieve dorsiflexion in the swing phase, because flexion of the hip and knee is accompanied by reflex dorsiflexion of the ankle.

Abnormal foot contact

The foot may be abnormally loaded in that the weight is primarily borne on only one of its four quadrants. Loading on the heel or forefoot is best

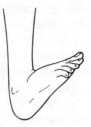

Fig. 3.15   *Talipes calcaneus.*

observed from the side, and loading on the medial or lateral side is best observed from the front, although some authorities state that the foot should always be observed from behind. Where it is possible, viewing from below the foot, using a glass walkway, gives an excellent idea of the pattern of foot loading.

Loading of the heel occurs in the deformity known as *talipes calcaneus* (also known as pes calcaneus), where the forefoot is pulled up into extreme dorsiflexion (Fig. 3.15), usually as a result of muscle imbalance such as results from spasticity of the anterior tibial muscles or weakness of the triceps surae. Except in mild cases, weight is never taken by the forefoot, and the stride length is reduced by the loss of the 'terminal rocker.'

In the deformity known as *talipes equinus* (or pes equinus) (Fig. 3.16), the forefoot is fixed in plantarflexion, usually through spasticity of the plantarflexors. In a mild equinus deformity, the foot may be placed onto the ground flat, but more commonly the heel never contacts the ground at all, and initial contact is made by the metatarsal heads, in a gait pattern known as *primary toestrike*. Because the line of force from the ground reaction is displaced anteriorly, an increased external moment tending to extend the knee is present. The loss of the initial rocker shortens the stride length.

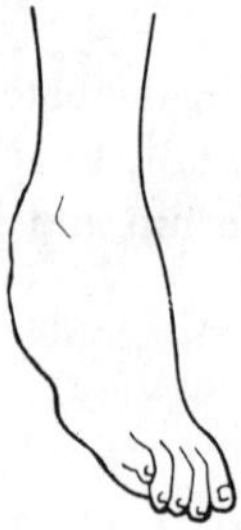

Fig. 3.16   *Talipes equinus.*

Excessive medial contact occurs in a number of foot deformities. Weakness of the invertors or spasticity of the evertors will cause the medial side of the foot to drop and to take most of the weight. In *pes valgus*, the medial arch is lowered, permitting weight-bearing on the medial border of the foot. Increased medial foot contact may also be due to a valgus deformity of the knee, accompanied by a widened walking base.

Excessive lateral foot contact may also result from a foot deformity where the medial border of the foot is elevated or the lateral border depressed by spasticity or weakness. The foot deformity known as *talipes equinovarus* (Fig. 3.17) combines equinus with varus, producing a curved foot where all the load is borne by the outer border of the forefoot. Although the term *club foot* may be applied to any foot deformity, it is most commonly used to mean talipes equinovarus.

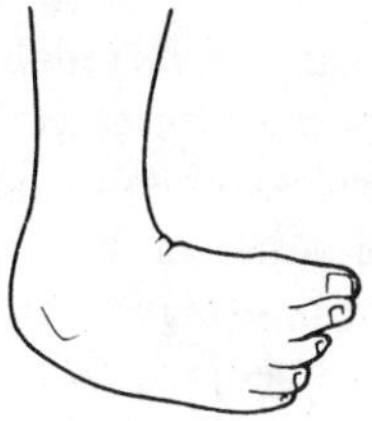

Fig. 3.17 *Talipes equinovarus.*

Another form of abnormal foot contact is the *stamping* that commonly accompanies a loss of sensation in the foot, such as occurs in tabes dorsalis. The subject receives feedback on ground contact from the vibration caused by the impact of the foot on the ground.

Insufficient push off

In normal walking, weight is borne on the forefoot during the 'push off' at the end of the stance phase. In the gait pattern known as insufficient push off, the weight is taken primarily on the heel, and there is no push off phase, the whole foot being lifted off the ground at once. It is best observed from the side.

The main cause of insufficient push off is a problem with the triceps surae or Achilles tendon, which prevents adequate weight-bearing on the forefoot. Rupture of the Achilles tendon, or weakness of the soleus and gastrocnemius, are typical causes. Weakness or paralysis of the intrinsic muscles of the foot may also prevent load from being taken through the forefoot.

Insufficient push off may also result from any foot deformity if the anatomy is so distorted that it prevents normal forefoot loading. A calcaneus deformity (Fig. 3.15) obviously makes it impossible to put any significant load on the forefoot.

Another important cause of insufficient push off is pain under the forefoot, if the amount of pain is affected by the degree of loading. This may occur in metatarsalgia, and arthritis affecting the metatarso-phalangeal joints. The loss of the terminal rocker causes the foot to leave the ground prematurely, before the hip has fully extended, which reduces the step length on the other side. The duration of the stance phase on the affected side is also reduced, which produces an asymmetry in gait timing.

Abnormal walking base

The walking base is usually in the range 50–100 mm. In pathological gait it may be either increased or decreased beyond this range. While ideally determined by actual measurement, changes in the walking base may be estimated by eye from either in front or behind.

An increased walking base may be caused by any deformity, such as an abducted hip or valgus knee, which causes the feet to be placed on the ground wide apart. Increased lateral movement of the trunk is required to maintain balance, as shown in Fig. 2.19.

The other important cause of an increased walking base is instability and fear of falling, the feet being placed wide apart to increase the area of support. This allows a margin of error in the positioning of the center of gravity over the feet. This gait abnormality is likely to be present when there is a deficiency in the sensation or proprioception of the legs, so that the subject is not quite sure where they are, relative to the trunk. It is also used in cerebellar ataxia, to increase the level of security in an uncoordinated gait pattern. In addition to the widened walking base, the use of one or two canes will considerably aid stability.

A narrow walking base usually results from an adduction deformity at the hip or a varus deformity at the knee. Hip adduction is commonly seen in cerebral palsy and the resulting gait pattern is known as *scissoring*. In milder cases, the swing phase leg is able to pass the stance phase leg, but lands on the ground in front of it. In more severe cases, the swing-ing leg is unable to pass the stance leg. After the leading leg has been advanced, the other leg is brought up behind it, on its lateral side, leading to a very short stride length. Under these circumstances, the walk-ing base will have a negative value. This is clearly a very disabling gait pattern.

Rhythmic disturbances

Gait disorders may include abnormalities in the timing of the gait cycle, as well as in the pattern of limb movement and limb loading. Two types of rhythmic disturbance can be identified: an *asymmetric* rhythmic disturbance, which shows a difference in the gait timing between the two legs; and an *irregular* rhythmic disturbance, which shows differences between one stride and the next. Rhythmic disturbances are best observed from the side, and may also be audible.

An *antalgic* gait pattern is specifically a gait modification designed to reduce the amount of pain a person is experiencing. The term is usually applied to a rhythmic disturbance in which as short a time as possible is spent on the painful limb and a correspondingly longer time is spent on the healthy side. The pattern is asymmetrical between the two legs but is generally regular from one cycle to the next. A marked difference in leg length between the two sides may also produce a regular gait asymmetry of this type, as may a number of other differences between the two sides, such as joint contractures or ankylosis.

Irregular gait rhythmic disturbances, where the timing alters from one step to the next, are seen in a number of neurological conditions. In particular, cerebellar ataxia leads to loss of the 'pattern generator' responsible for a regular, coordinated sequence of footsteps. Loss of sensation or proprioception may also cause an irregular arrhythmia, due to a general uncertainty about limb position and orientation.

Other gait abnormalities

A number of other gait abnormalities may be observed, either alone or in combination with some of the gait patterns described above. They include:

1. Abnormal movements, for example intention tremors and athetoid movements
2. Abnormal attitude or movements of the upper limb, including a failure to swing the arms
3. Abnormal attitude or movements of the head and neck
4. Sideways rotation of the foot following heelstrike
5. Rapid fatigue.

Walking aids

The use of walking aids may modify the gait pattern considerably. While some people use a walking aid to make it easier to walk, for example to reduce the pain in a painful joint, others are totally unable to walk without some form of aid. Although there are many detailed variations in design, walking aids can be classified into three basic types: *canes*, *crutches* and *frames*. All three operate by supporting part of the body weight through the arm rather than the leg. While this is an effective way of coping with inadequacies of the legs, it frequently leads to problems with the wrist and, especially, the shoulder joints, which are simply not designed for the transmission of large forces.

There is considerable variability in the way in which walking aids are used, and people will often use them in ways which do not quite fit the typical patterns described in the following sections.

Canes

The simplest form of walking aid is the cane, also known as a walking stick, by means of which force can be transmitted to the ground through the wrist and hand. Since the forearm muscles are relatively weak and the joints of the wrist fairly small, it is impossible to transmit large forces through a cane for any length of time. The torque which can be applied to the upper end of the cane is limited by the grip strength and by the shape of the handle, since the hand tends to slip. For this reason the major direction of force transmission is along the axis of the cane. Canes may be used for three purposes, which are often combined:

1. To improve stability
2. To generate a moment
3. To take part of the load away from one of the legs.

1. *Improve stability:* Canes are frequently used by the elderly and infirm to improve their stability. This is achieved by increasing the size of the area of support, thus removing the need to position the center of gravity over the relatively small supporting area provided by the feet. In those with only minor stability problems, a single cane may be used. This will not provide a secure supporting area during single limb support, but does make it easier to correct for small imbalances. Since the cane is usually placed on the ground some distance away from the feet, giving a relatively long lever arm, a modest force through it will produce a substantial moment to correct for any positioning error. For maximum security, a person will need to use two canes, so that a triangular supporting area is always available. This is provided by two canes and one foot during single limb support, and by one cane and two feet during double support. If only a single cane is used, it will normally be advanced during the stance phase of the more secure leg. If two canes are used,

they are usually advanced separately, during double limb support, to provide the maximum stability at all times.

2. *Generate a moment:* The use of a cane to generate a moment is illustrated in Fig. 3.18. This should be compared with Fig. 3.2, which shows the mechanics of normal single legged stance. A vertical force of 10 kgf is applied through the cane, which generates a counterclockwise moment applied to the shoulder girdle and hence to the pelvis. This reduces the size of the moment which the hip abductor muscles need to generate to keep the pelvis level. The contraction of these muscles is reduced by 40 kgf, and this reduces the total force in the hip joint by the same amount. For this mechanism to work, the cane must be held in the opposite hand to the painful hip. A cane may also be used to generate a lateral moment at the knee to reduce the loading on one side of the joint. The cane is advanced during the swing phase of the leg it is protecting.

3. *Reduce limb loading:* When using a cane to remove some of the load from the leg, it is usually held in the same hand as the affected leg and placed on the ground close to the foot. In this way, load-sharing can be achieved between the leg and the cane, even to the extent of removing the

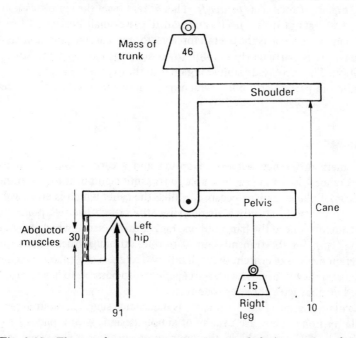

Fig. 3.18 *The use of a cane to generate a counterclockwise moment reduces the contraction force of the hip abductors, and hence the force in the hip joint (see note to* Fig. 3.1).

load entirely from the leg. The cane follows the movements of the affected leg, being advanced during the swing phase on that side. The person will normally lean sideways over the cane, in a *lateral lurch*, to increase the vertical loading on it, and hence to reduce the load in the leg. A cane may be used in this way to relieve pain in the hip, knee, ankle or foot. If the cane is held in the opposite hand, as is often recommended, this lateral lurching can be avoided, but the degree of offloading is reduced.

Whichever of these three reasons the person has for using a cane, the degree of disability will determine whether one or two canes are used. Sometimes it may be observed that a subject uses a cane in the opposite hand from what might have been expected. In some cases he or she has simply not discovered that they would benefit from using the cane in the other hand, but more often the observer has failed to appreciate fully all the compensations which the subject has adopted.

There are a number of ways in which the simple cane can be modified, including many different types of handgrip. A particularly important variant on the simple cane is the *broad based cane*, which may have three feet (*tripod*) or four feet (*tetrapod*). This differs from the simple cane in that it will stand up by itself, and will tolerate small horizontal force components as long as the overall force vector remains within the area of its base. It is particularly helpful when standing up from the sitting position. The increased stability is gained at the expense of an increase in weight and particularly bulk, which may cause difficulties when going through doorways.

Crutches

The main difference between a crutch and a cane – apart from its appearance – is that a crutch is able to transmit significant forces in the horizontal plane. This is because, unlike the cane, which is effectively fixed to the body at only a single point, the crutch has two points of attachment – one at the hand and one higher up the arm – which provide a lever arm for the transmission of torque. Although there are many different designs of crutch, they fall into two categories: axillary crutches and forearm crutches. As with the cane, it is also possible to have a broad based crutch, with three or four feet.

Axillary crutches (Fig. 3.19, left), as their name suggests, fit under the axilla (armpit). They are usually of simple design, with a padded top surface and a handhold in the appropriate position. The lever arm between the axilla and the hand is fairly long, and enough horizontal force can be generated to permit walking when both legs are straight and

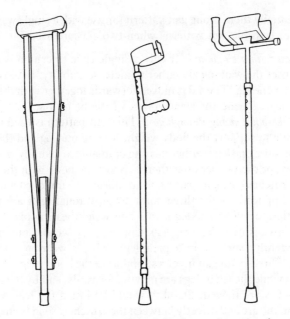

Fig. 3.19   *Three types of crutch: axillary (left); forearm (center) and gutter (right).*

non-functional. A disadvantage of this type of crutch is that the axilla is not an ideal area for weightbearing, and incorrect fitting or prolonged use may cause damage to the blood vessels or nerves. Although some people use axillary crutches for many years, they are more suitable for short term use, for instance while a patient has a broken leg set in plaster.

There are many different types of *forearm crutches*, also called *elbow crutches* (Fig. 3.19, center). They differ from axillary crutches in that the upper point of contact between the body and the crutch is provided by either the forearm or the upper arm, rather than by the axilla. The lever arm is thus shorter than for an axillary crutch, although this is seldom a problem, and they usually run less risk of tissue damage as well as being lighter and more acceptable cosmetically. In the normal forearm crutch, most of the vertical force is transmitted through the hand, but the use of a 'gutter' or 'platform' permits more load to be taken by the forearm itself (Fig. 3.19, right).

Gait patterns with walking aids

There are a number of different ways of walking when using a cane or crutch. The terminology varies somewhat from one author to another, but essentially there is one gait pattern for use with a single walking aid, and three possible gait patterns when two aids are used.

1. *Gait with a single aid:* If only a single cane or crutch is used, the subject uses the walking aid either to increase stability, or to reduce the loading on one leg. The aid is moved forwards together with the worse of the two legs, during the stance phase of the better one.

2. *Three-point swing-through gait:* This gait pattern is used when it is impossible to support the body weight first on one leg and then on the other, because one leg is either missing or unable to take any weight. The single leg, or the two legs together, support the body with the help of a pair of crutches (the gait pattern is not normally employed with canes). Typical applications for three-point swing-through gait are following amputation, or when one leg is in a non-weightbearing plaster. It may also be used when the two legs can support weight but not move independently, such as in a paraplegic who is wearing a full-length calliper. Three-point gait involves support of the body weight by the two crutches while the leg or legs are moving forwards, and by the legs while the crutches are moved. Stability would be lost if the legs were to be placed on the ground directly between the crutches, so in swing-through gait they are brought up in front of the level of the crutches. Figure 3.20 shows three-point swing-through gait in a person taking weight on both legs.

3. *Three-point swing-to gait:* This gait pattern is similar to three-point swing-through gait, except that the feet are advanced by a much shorter distance, being placed on the ground behind the level of the crutches. The crutches are then advanced, and the feet brought up behind them

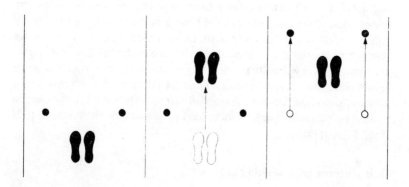

Fig. 3.20 *Three-point swing-through gait in a person taking weight on both legs. The legs are advanced together, to in front of the line of the crutches.*

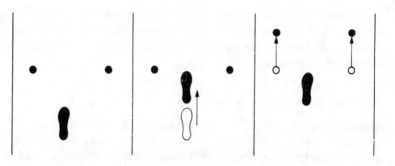

Fig. 3.21   *Three-point swing-to gait in someone taking weight on only one leg. The leg is advanced to behind the line of the crutches.*

again. Figure 3.21 shows this pattern in a person able to bear weight on only one foot. Swing-to gait involves a much shorter stride length, and is less efficient, than swing-through gait. Which of the two patterns is adopted depends on factors such as the subject's stability and upper limb strength. Both swing-to and swing-through gait involve large vertical excursions of the body's center of gravity, and are costly in terms of energy usage.

4. *Four-point gait:* As shown in Fig. 3.22, four-point gait involves an essentially normal walking action by the legs, with extra help being given by two canes or crutches. It is thus only appropriate when both legs are able to support part of the body weight. Subjects who have only minor stability problems may use two canes, each of which is moved forwards during the swing phase of the opposite leg, during which time the body has only two points of contact with the ground. Those who are less stable will probably use crutches, and will move them only when the other

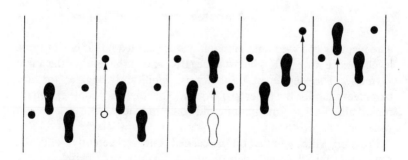

Fig. 3.22   *Four-point gait. One crutch or leg is moved at a time, in the pattern: left crutch – right leg – right crutch – left leg.*

crutch and both legs are supporting the body, thus ensuring three points of contact with the ground at all times. People differ in how the canes or crutches are advanced, but the pattern of Fig. 3.22 is typical: left crutch – right leg – right crutch – left leg.

Walking frames and rollators

The most stable walking aid is the frame, also called a 'walker' or 'zimmer frame,' which enables the subject to stand and walk within the area of support provided by its base. Considerable force can be applied to the frame vertically, and moderate forces can be applied horizontally, provided that the overall force vector remains within the area of support. The usual method of walking is first to move the frame forwards, then to take a short step with each foot, then to move the frame again, and so on. Walking is thus extremely slow, with a start–stop pattern. Although the subject is encouraged to lift the frame forward at each step, it is often simply pushed along the ground.

A *rollator* is a variant on the walking frame, in which the front feet are replaced by wheels. This makes it easier to advance, at the expense of a slight reduction in stability in the direction of progression. The mode of walking is very similar to that with the frame, except that it is easier to move forwards, since tipping the rollator lifts the back feet clear of the ground. Again, many subjects misuse the rollator by sliding it forwards, rather than tipping it. A further variant on the design is to replace all four feet by wheels, which works well if effective brakes are provided. There are many other designs of frames and rollators, including ones which fold, and ones with gutters to support the body weight through the forearms.

Amputee gait

The majority of lower limb amputations are carried out at one of three levels: above knee (AK), below knee (BK) and at the level of the ankle (Syme's). The degree of disability is significantly greater for AK amputees than for the other two groups. Nonetheless, people can adapt remarkably well to amputation, and some individuals are able to run using an AK prosthesis.

Providing the amputation is carried out competently, and the remaining muscles and nerves are normal, people with an AK amputation should have useful function in all the muscle groups acting about the hip joint. However, no matter how well the prosthetic socket is fitted, the mechanical coupling between the femur and the prosthetic limb can never be as good as in the normal individual, for three reasons:

1. The lever arm between the hip joint and the socket is relatively small, which reduces the moment which the hip muscles can apply to the prosthetic leg.
2. There is always some relative motion between the stump of the femur and the socket, due to the compression of soft tissues, which is exaggerated in the case of a poorly fitting socket.
3. If the socket is uncomfortable, the subject may be reluctant to apply large forces to the prosthetic limb.

AK amputees lack any ability to resist an external flexion moment at the knee, and they must walk with the knee in full extension. They commonly use the gait modification of anterior trunk bending in order to keep the knee fully extended. Walking with a stiff knee leads to greater rise and fall of the center of gravity than normal, and a consequent increase in energy expenditure. This type of walking is tiring, but nonetheless, some people with one or even two AK amputations manage to get around extremely well. There have been some attempts to provide stance phase knee flexion, or its equivalent (shortening the leg through a telescopic action), but these devices have not come into widespread use.

Control of the knee joint in the swing phase is one of the most important requirements in the design of an AK prosthesis. If the knee joint is completely free, the pendulum action causes it to flex too quickly following toe off, so that the heel flicks up behind the subject. The knee then extends too rapidly, stopping abruptly as it hits the end stop in hyperextension. In contrast, a knee which permits no flexion or extension requires the whole leg to be accelerated and decelerated in one piece, which places enormous demands on the hip musculature, and leads to much higher energy expenditure in walking (Saunders et al., 1953). The compromise between these extremes is a prosthetic knee joint with some form of damping mechanism, which thus prevents all of these problems.

A variety of different damping mechanisms are available, including friction, hydraulic and pneumatic systems. Murray et al. (1983) examined the gait of AK amputees using two different knee mechanisms, one of which gave a constant frictional load at all knee angles, and one, using hydraulics, which varied its loading with knee angle and direction of motion. They found that the performance of the hydraulic mechanism was generally better. Amongst the differences they noted between normal and AK amputee gait, the duration of the swing phase was found to be longer on the amputated side, which resulted in a decreased cadence. The stride length was close to normal, but the decreased cadence led to a decreased velocity. Lateral trunk bending was often present; they attributed this partly to a compensation for a wider walking base, which was used to improve stability, and partly to a compensation for a decreased efficiency of the hip abductors, due to movement of the femoral stump within the socket. Heel rise took place earlier in the stance phase than in normal individuals, because of a reduction in the ability to dorsiflex the ankle. The magnitude of the heel rise in early swing was increased, especially at higher walking speeds – it is very sensitive to the frictional properties of the knee mechanism. There was a tendency to vault on the normal leg during the swing phase of the prosthetic leg. Since the prosthetic leg shortened adequately during the swing phase, this vaulting was probably not necessary. It may have been used to gain added security, since the ground clearance of the prosthetic limb cannot be judged in the absence of proprioception.

Figure 3.23 shows plots of the hip, knee and ankle angles in a 17-year-old female AK amputee with a hydraulic knee mechanism. It should be compared with Fig. 2.5, which shows comparable data from a normal subject. The knee moves into a few degrees of hyperextension before the

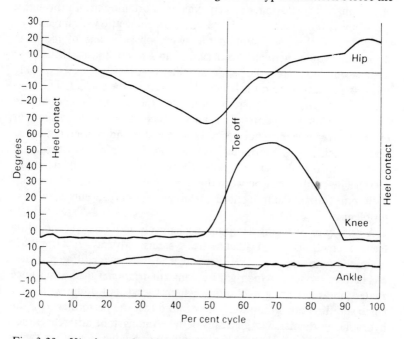

Fig. 3.23   *Hip, knee and ankle angles during walking in an above knee amputee. The main abnormality is hyperextension of the knee (cf Fig. 2.5).*

end of the swing phase, and remains hyperextended until near the end of the stance phase, which is a little shorter than normal. The swing phase knee flexion is almost normal. The hip movement is also almost normal, except for a sudden increase in flexion late in the swing phase, as the knee mechanism reaches its extension stop. Flexibility in the foot mechanism leads to a normal pattern of ankle motion, although the magnitudes of these movements are less than normal.

A BK amputation deprives the user of the ability to plantarflex and dorsiflex the ankle, although the artificial foot normally provides some flexibility in this axis, and its shape provides partially functioning initial and terminal rockers. The loss of active plantarflexion at the end of the stance phase means that muscle power cannot be used to provide an active 'push off,' and the effective length of the leg is shorter than normal, so that it has to be lifted clear of the ground sooner (Breakey, 1976). According to Saunders et al. (1953), the path of the center of gravity is essentially normal after a BK amputation, since the hip and knee together can largely compensate for the loss of the ankle joint. If both ankle and knee are lost, as in an AK amputation, the compensation is incomplete.

Figure 3.24 shows the hip, knee and ankle angles of a 47-year-old male

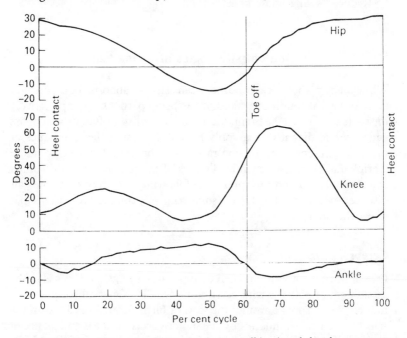

Fig. 3.24 *Hip, knee and ankle angles during walking in a below knee amputee. The ankle angle shows slight abnormalities (cf Fig. 2.5).*

BK amputee using a 'multiflex' foot, walking with a cadence of 95 steps/min, stride length 1.45 m and velocity 1.15 m/s, all of which are towards the lower end of the normal range. The hip and knee angles are entirely normal. The ankle angle is also just within normal limits, although the movement into plantarflexion at the end of the stance phase is of relatively low magnitude and occurs a little late. This is because it is a passive movement, resulting from the removal of loading from the elastic foot mechanism, rather than the active plantarflexion seen in normal subjects.

There are considerable differences between subjects in amputee gait. The subject whose gait is illustrated in Fig. 3.24 walks almost normally. However, according to Breakey (1976), the 'typical' gait pattern of the BK amputee includes:

1. Delayed foot flat
2. Reduced stance phase knee flexion
3. Early heel off
4. Early toe off
5. Reduced stance phase duration
6. Reduced swing phase knee flexion.

Common pathologies affecting gait

While many pathological conditions can cause an abnormal gait, a group of neurological conditions stand out as being particularly important. If the basic defect is in the brain, the gait abnormality is often very complex and accurate diagnosis may only be possible using the techniques of modern gait analysis. In contrast, gait abnormalities due to more 'peripheral' disorders, such as diseases of the joints, tend to be much easier to identify and interpret. The following sections outline the gait disorders which may result from some relatively common conditions that affect the brain.

Cerebral palsy

One of the most important current applications of gait analysis is in the assessment of patients with cerebral palsy (CP). Cerebral palsy usually follows brain damage around the time of birth. It involves a loss of the selective control of muscles by the motor cortex, and the emergence of spasticity and primitive patterns of contraction. As a general rule, the muscles are not weak, but they may be unable to contract adequately at the appropriate times in the gait cycle, due to a loss of coordination. Muscular contraction cannot be turned on or off rapidly, and there is commonly a co-contraction of antagonists.

The characteristics of CP are given by Gage as follows:

1. Abnormal tone which varies with position and/or movement
2. In a growing child, a propensity to develop muscle contractures
3. Loss of selective control of muscles
4. The necessity to use primitive reflexes to accomplish ambulation
5. Difficulty with balance.

There is considerable variation between one patient and another in the way in which cerebral palsy affects the positions and movements of the joints. The clinical picture depends on which muscles are affected, and on the timing of their contraction during the gait cycle. Since the neurological deficit is permanent and irreparable, the timing of muscular contraction cannot be altered by treatment. However, four basic methods of treatment can be employed:

1. Muscles can be made stronger by training
2. Muscles can be made weaker or non-functional by cutting or lengthening their tendons
3. The mode of action of muscles can be changed by tendon transplantation
4. Orthoses can be used to limit the movement of a joint, or to apply a force in a particular direction.

The two main varieties of cerebral palsy which affect the gait are spastic hemiplegia and spastic diplegia. The use of gait analysis to plan and monitor the treatment of cerebral palsy is discussed further in Chapter 5.

Spastic hemiplegia

Spastic hemiplegia is the commonest neurological cause of an abnormal gait. As well as occurring in cerebral palsy, it is also frequently seen in elderly people who have suffered a cerebrovascular accident ('stroke'). It may also occur following traumatic brain damage. It is characterized by spasticity and loss of function in some or all of the muscles of one leg, the other leg being normal or nearly normal. A hemiplegic patient usually possesses a mixture of normal motor control, spasticity and 'patterned responses,' the exact combination depending on the severity and location of the brain damage (Perry, 1969). As well as problems with moving and controlling their limbs, many hemiplegic patients also experience difficulty in maintaining balance, because a defect in the 'body image' causes them to ignore the affected side. Winters et al. (1987) showed that the gait pattern of children and young adults with this condition could be divided into four groups. These are numbered I to IV, in order of increasing severity, each group having all the neurological deficits of the preceding one, with some addition. In general, the results of this study of young people agreed with other publications on the gait of elderly subjects who had suffered a cerebrovascular accident.

Group I subjects essentially suffer from a single problem – a foot drop on the affected side. This causes initial contact to be by primary toestrike, and also produces a functional increase in leg length during the swing phase. According to Winters et al. (1987), the plantarflexed attitude of the foot at initial contact leads to an increased flexion of both the knee and hip. The lumbar lordosis is also increased throughout the gait cycle. The most common surgical treatment for this problem is to lengthen the Achilles tendon. However, this is unlikely to improve the gait, since the ankle is already able to dorsiflex adequately. The significant problem in these individuals is the foot drop during the swing phase, which is caused by a relative weakness of the anterior tibial muscles, and which can be treated adequately by a simple orthosis.

Group II subjects, as well as having a foot drop, have a static or dynamic contracture of the calf muscles, which holds the ankle in plantarflexion throughout the whole of the gait cycle. The difference in gait between Group I and Group II subjects is seen after mid stance, when the persistent plantarflexion produces an external moment which forces the knee into hyperextension. According to Winters et al. (1987), advancement of the trunk is curtailed, and the length of the opposite step is decreased. As with the Group I subjects, there is increased hip flexion with an associated increased lumbar lordosis. These patients generally benefit from an operation to lengthen the Achilles tendon, as well as an orthosis to control the foot drop.

Group III subjects have a foot drop, contracted calf muscles, and overactivity of both quadriceps and hamstrings. This causes a reduction in the total range of motion at the knee, with a marked reduction in swing phase flexion. The gait is described as 'stiff with short steps.' The other features of the Group II pattern are also present: hyperextension of the knee in late stance, hip flexion and increased lumbar lordosis. Surgical correction of these patients will extend as high as the knee. Waters et al. (1979) showed that the stiff-legged gait is caused by an inappropriate contraction of one or more heads of the quadriceps at the end of the stance phase and the beginning of the swing phase. They described the surgical treatment of the condition by quadriceps tenotomy.

The greatest degree of neurological involvement is seen in Group IV subjects. In addition to the characteristics of the Group III subjects, they also have a reduced range of hip motion, due to overactivity of iliopsoas and the adductors. The hip is unable to extend fully, so anterior pelvic tilt and increased lumbar lordosis at the end of the stance phase are used to preserve the stride length. Full treatment of these patients will include surgery to the muscles acting about the hip, knee and ankle joints.

Spastic diplegia

Spastic diplegia is a common manifestation of cerebral palsy. It affects both legs although there may be considerable asymmetry between the two sides. There are considerable variations between individuals, but the commonest pattern consists of:

1. Hip flexion and internal rotation, due to overactivity by the iliopsoas, rectus femoris and hip adductors
2. Knee flexion, due to overactivity of the hamstrings, especially on the medial side
3. Equinus deformity of the foot and eversion of the hindfoot, due to overactivity by the triceps surae and peronei.

A less common pattern includes hyperextension of the knee and a 'stiff-legged gait,' which interferes with foot clearance during the swing phase.

In individuals suffering from the commonest pattern of spastic diplegia, excessive flexion of the hips leads to an increased lumbar lordosis in order to get the femur as vertical as possible (see Fig. 3.8). The knee is held almost fixed by co-contraction of the hamstrings and quadriceps. Since the hamstrings are the more powerful, the knee remains flexed, and its angle typically varies only between 30 and 40 degrees. The equinus deformity of the foot causes a primary toestrike, and loss of all three 'rockers' – plantarflexion following heelstrike, dorsiflexion in mid stance and plantarflexion in late stance.

A typical case of spastic diplegia is shown in Fig. 3.25 (adapted from Gage, 1983), which shows the hip, knee and ankle angles on the left side of a six-year-old girl with this condition. The hip angle, defined here as the angle between the femur and vertical, was normal, although the pelvis was tilted forwards to compensate for tight hip flexors. The knee never extended beyond 20 degrees of flexion, due to overactivity of the hamstrings. The dorsiflexed position of the ankle throughout the cycle is evident, initial contact being made by the forefoot rather than the heel because of spasticity of the triceps surae. Particularly striking on this plot is the very long stance phase and short swing phase.

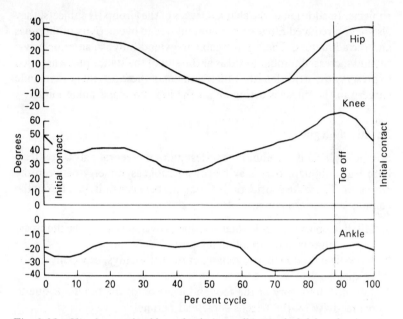

Fig. 3.25   *Hip, knee and ankle angles during walking in the left leg of a six-year-old girl with spastic diplegia (adapted from Gage, 1983) (cf Fig. 2.5).*

An occasional complication of the treatment of cerebral palsy is the development of a *crouch gait* (Fig. 3.26), which may occur if an equinus deformity of the foot is overcorrected in the presence of a flexion contracture of the knee (Sutherland and Cooper, 1978). Excessive lengthening of the Achilles tendon may be caused directly by surgery, or it may be caused by postoperative stretching. The result is a calcaneus deformity of the foot, which keeps the line of the ground reaction force a long way behind the already flexed knee. This leads to a vicious circle in which the increased external moment leads to an increased force in the quadriceps, which gradually stretches the patellar tendon, further increasing the knee angle and hence the external moment. Sutherland and Cooper stressed the need to correct the knee deformity before operating on the heel cord.

Crouch gait may also occur in cerebral palsy without any operative intervention, as a result of spasticity of either the hip flexors or the hamstrings.

Parkinsonism

The condition known as parkinsonism is a disorder of the extra-

Fig. 3.26 *Position of body at mid stance in a 12-year-old child with crouch gait, following Achilles tendon lengthening for spastic diplegia (adapted from Sutherland and Cooper, 1978).*

pyramidal system, caused by degeneration of the basal ganglia of the brain. Amongst its other clinical features, it includes a *shuffling gait*. Murray et al. (1978) examined the gait of 44 men with parkinsonism, and identified the following abnormalities:

1. Stride length and velocity were very much reduced, although cadence was usually normal
2. The walking base was slightly increased
3. The range of motion at the hip, knee and ankle were all decreased, mainly by a reduction in the amount of joint extension
4. The swinging of the arms was much reduced
5. The majority of patients rotated the trunk in phase with the pelvis, instead of twisting it in the opposite direction
6. The vertical trajectories of the head, heel and toe were all reduced, although other workers have commented on a distinct 'bobbing' motion of the head.

The 'shuffle' occurred because the foot was still moving forward at the time of initial foot contact. In some patients initial contact was made by the flat foot; in others there was a heelstrike, but the foot was much more horizontal than usual. Some patients also showed scuffing of the foot in mid swing. Unlike most gait patterns, which stabilize within the first two or three steps, the gait of patients with parkinsonism often 'evolves' over the course of several strides.

4

Methods of Gait Analysis

When considering the methods which may be used to perform gait analysis, it is useful to regard them as being in a 'spectrum' or 'continuum,' ranging from the absence of technological aids at one extreme, to the use of complicated and expensive equipment at the other. This chapter starts with a method which requires no equipment at all, and goes on to describe progressively more elaborate systems. As a general rule, the more elaborate the system, the higher the cost, but the better the quality of objective data that can be provided. However, this does not imply that some of the simpler techniques are not worth using. It has often been found, particularly in a clinical setting, that the use of high-technology gait analysis is inappropriate, because of its large cost in terms of money, space and time, and because the clinical problem can be adequately managed using simpler techniques.

Appendix 4 lists all the manufacturers of gait analysis equipment that the author has been able to discover.

Visual gait analysis

It is tempting to say that the simplest form of gait analysis is that made by the unaided human eye. This, of course, neglects the remarkable abilities of the human brain to process the data received by the eye. Visual gait analysis is, in reality, the most complicated and versatile form of analysis available. Despite this, it suffers from four serious limitations:

1. It is transitory, giving no permanent record
2. The eye cannot observe high-speed events
3. It is only possible to observe movements, not forces
4. It depends entirely on the skill (or luck) of the individual observer.

In a study on the reproducibility of visual gait analysis, Krebs et al. (1985) found it to be 'only moderately reliable.' Saleh and Murdoch (1985) compared the performance of people skilled in visual gait analysis with the data provided by a combined kinetic/kinematic system. They found that the measurement system identified many more gait abnormalities than had been seen by the observers.

Many clinicians include the observation of a subject's gait as part of their clinical examination. Whilst this is a laudable practice, it will not provide an adequate gait analysis if it is simply limited to watching the subject walk up and down the room once. This merely gives a superficial idea of how well they walk, and perhaps identifies the most serious abnormality. A thorough visual gait analysis, as recommended in the manual from New York University (1986), involves the subject making a number of walks, which are observed from each side, from the front, and from the back. As the subject walks, the observer should look for the presence or absence of a number of specific gait abnormalities, such as those described in Chapter 3 and summarized in Table 4.1. A logical order should be used for looking for the different gait abnormalities – the mixture of walking directions listed in the table is not recommended! According to Rose (1983), it is also important, when performing visual gait analysis, to compare the range of motion at the joints during walking with those that are observed on the examination table – they may be either more or less.

The minimum length required for a gait analysis walkway is a hotly debated subject. The author believes that 8 m (26 ft) is about the

Table 4.1 *Common gait abnormalities and best viewpoint for observation.*

Gait abnormality	Observing direction
Lateral trunk bending	Side
Anterior trunk bending	Side
Posterior trunk bending	Side
Increased lumbar lordosis	Side
Circumduction	Front or behind
Hip hiking	Front or behind
Steppage	Side
Vaulting	Side or front
Abnormal hip rotation	Front or behind
Excessive knee extension	Side
Excessive knee flexion	Side
Inadequate dorsiflexion control	Side
Abnormal foot contact	Front or behind
Insufficient push off	Side
Abnormal walking base	Front or behind
Rhythmic disturbances	Side

minimum for use with fit young people, but that 10–12 m (33–39 ft) is preferable, since it permits fast walkers to 'get into their stride' before any measurements are made. However, shorter walkways are satisfactory for people who walk more slowly. This particularly applies to those with a pathological gait, since the gait pattern usually stabilizes within the first two or three steps. A notable exception to this, however, is the gait in parkinsonism, which 'evolves' over the first few strides. The width required for a walkway depends on what equipment, if any, is being used to make measurements. For visual gait analysis, as little as 3 m may be sufficient. If videotape is being used, the camera needs to be positioned a little further from the subject, and about 4 m is needed. A kinematic system making simultaneous measurements from both sides of the body normally requires at least 5–6 m (16–20 ft). Figure 4.1 shows the layout of a small gait laboratory used for visual gait analysis, videotape and the measurement of the general gait parameters.

Some investigators permit subjects to choose their own walking speed, whereas others control the cadence, for example by asking them to walk in time with a metronome. The rationale for controlling the cadence is that many of the measurable parameters of gait vary with the walking speed, and using a controlled cadence provides one means of reducing the variability. However, subjects are unlikely to walk naturally under these conditions, and patients with motor control problems may find it difficult or even impossible to walk at an imposed cadence. The answer to this dilemma is probably to accept the fact that subjects need to walk at different speeds, and to interpret the data appropriately. This means that

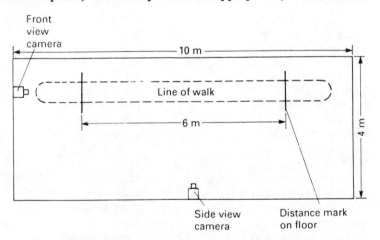

Fig. 4.1 *Layout of a small gait laboratory used for visual gait analysis, videotaping and measurement of the general gait parameters.*

'normal' values must be available for a range of walking speeds. However, an unresolved difficulty with this approach is that it may not be possible to get 'normal' values for very slow walking speeds, since normal individuals do not customarily walk very slowly, and when asked to do so, the gait parameters may become very variable (Brandstater et al., 1983).

Gait assessment

Simply observing the gait and noting abnormalities is of little value by itself. This needs to be followed by gait assessment, which is the synthesis of these observations with other information about the subject, obtained from the history and physical examination, combined with the intelligence and experience of the observer (Rose, 1983). Visual gait analysis is entirely subjective, and the quality of the analysis depends on the skill of the person performing it. It can be an interesting exercise to perform visual gait analysis on people walking past in the street, but without knowing the clinical details, it is easy to misinterpret an abnormality completely.

One thing that must constantly be borne in mind, when performing any type of gait analysis is that you are observing effects and not causes. Putting it another way, the observed gait pattern is not the direct result of a pathological process, but the net result of a pathological process and the subject's attempts to compensate for it. The observed gait pattern is what is left after the available mechanisms for compensation have been exhausted (Rose, 1983).

Videotape examination

The video cassette recorder (VCR) is one of the most useful pieces of equipment to be introduced into gait analysis during recent years. It helps to overcome two of the limitations of visual gait analysis – the lack of a permanent record and the difficulty of observing high speed events. In addition, it confers the following advantages:

1. It reduces the number of walks a subject needs to do
2. It makes it possible to show subjects exactly how they are walking
3. It makes it easier to teach visual gait analysis to someone else.

Videotape analysis is not an objective method, since it does not provide quantitative data in the form of numbers. However, it does provide a permanent record which can be extremely valuable. The presence of an earlier videotape of a subject's gait may be used to demonstrate to all concerned how much progress has been made, especially when this has occurred over a long period of time. In particular, it may convince a subject or family member that an improvement *has* occurred, when memory tells them that they are no better than they were several months ago!

Providing it has good slow motion and stop-frame facilities, videotape may be used to visualize events that are too fast for the unaided eye. While such features used to be found only on expensive commercial systems, they are now available on many domestic VCRs, some of which are quite suitable for use in a clinical gait laboratory. The most practical system consists of a camera-recorder ('camcorder'), to do the videotaping, and a separate VCR to replay the tapes. Highly desirable features on the camcorder are a zoom lens, automatic focus and the ability to operate in normal room lighting. A solid state charge-coupled device (CCD) camera is also a considerable advantage, since it eliminates blurring due to movement. The VCR used to view the tapes should have a rock-steady freeze-frame facility, without any 'stripes' across the picture, and the ability to single-step successive frames, either one at a time or at a very slow speed.

In making a thorough visual gait analysis without the use of videotape, the subject needs to make repeated walks to confirm or refute the presence of each of the gait abnormalities listed in Table 4.1. If the subject is in pain or easily fatigued, this may be an unreasonable requirement, and it may be difficult to achieve a satisfactory analysis. The use of videotape permits the subject to do a much smaller number of walks, as the person performing the analysis can rewind the tape and watch it as many times as necessary.

Videotape facilitates the process of teaching visual gait analysis, in which the student often needs to see small abnormal movements which happen very quickly. It is much easier to see such movements if the gait can be examined in slow motion, with the instructor pointing out details on the television monitor. The use of videotape also makes it possible to observe a variety of abnormal gaits which have been archived on tape.

Showing the subject a videotape of their own gait is not exactly 'biofeedback', since there is a time delay involved, but nonetheless it can be very helpful. When a therapist is working with a subject to correct a gait abnormality, the subject may gain a clearer idea of exactly what the therapist is concerned about if they can observe their own gait from an external viewpoint.

Although visual gait analysis using videotape is subjective, it is easy, at the same time, to derive some objective data: the general gait parameters (cadence, stride length and velocity). The method will be described in the next section. It is also possible to measure joint angles, either directly from the monitor screen, or using some form of on-screen digitizer. Such measurements tend to be fairly inaccurate, however, because the limb may not be viewed from the correct angle, and because distortions may be introduced by the television camera, the VCR and the television display.

Individual investigators will find their own ways of performing gait analysis using videotape, but the author's own routine may prove useful as a starting point. Subjects are asked to wear shorts or a swimsuit, so that the majority of the leg is visible. It is important that the subject should walk as 'normally' as possible, so they are asked to wear their own indoor or outdoor shoes, with socks if preferred. Surprisingly, subjects are seldom troubled by being videotaped while wearing this strange combination of clothing! Unless it would unduly tire the subject, it is a good idea to make one or two practice walks before starting videotaping. The camera position, or the zoom lens, is first adjusted to show the whole body from head to feet, and the subject is videotaped from the side, as they walk the length of the walkway in one direction. At the end they turn around, with a rest if necessary, and are videotaped as they walk back again. The whole process is then repeated, with the camera adjusted to show a close-up of the body from the waist down. Then either the camera position is changed, or the subject is asked to walk along a different pathway, so that they can be videotaped while walking away from the camera and then back towards it again. Two walks in each direction are again videotaped, although the magnification does not need to be changed, since this view automatically provides a range of views from close-up to full-body.

Subjects should not be able to see themselves on a monitor while they are walking, as this provides a distraction, particularly for children. Whether they are shown the videotape afterwards is at the discretion of the investigator, although it is important to review the tape before the subject leaves, in case the recording needs to be repeated for some reason.

The analysis is performed by replaying the videotape, looking for specific gait abnormalities in the different views, and interpreting what is seen in the light of the subject's history and physical examination. It is particularly helpful if two or more people work together to perform the analysis. Rose (1983) suggested that gait analysis should be based on the team approach, with discussion and hypothesis testing. As will be described in Chapter 5, hypothesis testing may involve an attempted modification of the gait, for instance by fitting an orthosis, or by paralyzing a muscle using local anesthetic.

General gait parameters

The general gait parameters are the cadence, stride length and velocity. These provide the simplest form of objective gait evaluation (Robinson and Smidt, 1981). Although there are automatic ways of making these measurements, which will be described in the next section, they may also be made using only a stopwatch, a tape measure, and some talcum powder.

Cadence, stride length and velocity tend to change together in most locomotor disabilities. For example, a subject with a slow cadence will usually also have a short stride length and a low velocity (velocity being the product of cadence and stride length). The general gait parameters give a guide to the walking ability of a subject, but little specific information. They should always be interpreted in terms of the expected values for the subject's age and sex (see Appendix 1). Figure 4.2 shows one way in which these data may be presented; the diamonds represent the 95 per cent confidence limits for a normal subject of the same age and sex as the subject under investigation.

Many workers in the field tend to 'normalize' the general gait parameters, and also other parameters such as joint moments, using arthropometric parameters such as height and weight. The author does

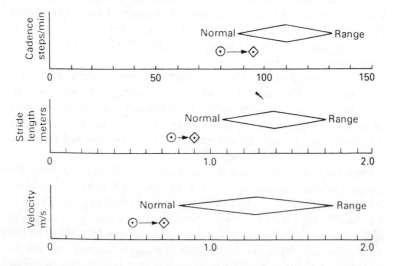

Fig. 4.2   *Display of the general gait parameters, with normal ranges appropriate for a patient's age and sex. Pre- and postoperative values for a patient undergoing knee replacement surgery.*

not follow this practice, since the improvements in accuracy which result are only small, the units become awkward (e.g. statures per second) and it is often impossible to make similar calculations on published data.

Cadence

Cadence may be measured with the aid of a stopwatch, by counting the number of individual steps taken during a known period of time. It is seldom practical to count for a full minute, so a period of 10 or 15 seconds is usually chosen. The loss of accuracy incurred by counting for such short periods of time is unlikely to be of any practical significance. The subject should be told to walk naturally, and they should be allowed to get up to their full walking speed before the observer starts to count the steps. The cadence is calculated using the formula:

$$\text{cadence (steps/min)} = \text{steps counted} \times 60/\text{time (s)}$$

Stride length

Stride length can be determined in two ways—by direct measurement, or indirectly from the velocity and cadence. The simplest direct method of measurement is to count the strides taken while the subject covers a known distance. A more useful method (but a rather messy one!) is where the subject steps with both feet in a shallow tray of talcum powder, and then walks across a polished floor, leaving a trail of footprints. These may be measured, as shown in Fig. 2.3, to derive left and right step lengths, stride length, walking base, toe-out angle, and some idea of the foot contact pattern. This investigation is able to provide a great deal of useful information, for the sake of a few minutes of mopping up the floor afterwards! As an alternative to using talcum powder, felt adhesive pads, soaked in different colored dyes, may be fixed to the feet (Rose, 1983). The subject walks along a strip of paper, and leaves a pattern of dots which give an accurate indication of the locations of both feet.

If both the cadence and the velocity have been measured, stride length may be calculated using the formula:

$$\text{stride length (m)} = \text{velocity (m/s)} \times 120/\text{cadence (steps/min)}$$

As was explained in Chapter 2, the number 120 is used because the cadence is a measure of half-strides, not full strides. For accurate results, the cadence and velocity should be measured during the same walk. However, the simultaneous counting, measuring and timing may prove to be too difficult, and the errors introduced by using data from different walks are not likely to be important unless the subject's gait varies markedly from one walk to another.

Velocity

The velocity may be measured by timing the subject while he or she walks a known distance, for example between two marks on the floor, or between two pillars in a corridor. The distance walked is a matter of convenience, but somewhere in the region of 6–10 m (20–33 ft) is probably adequate. Again, the subject should be told to walk at their natural speed, and they should be allowed to get into their stride before the measurement starts. The velocity is calculated as follows:

$$\text{velocity (m/s)} = \text{distance (m)/time (s)}$$

General gait parameters from videotape

Determination of the general gait parameters from a videotape of the subject walking is perfectly simple, providing the tape shows the subject passing two landmarks whose positions are known. One simple method is to make two lines on the floor, a known distance apart, using adhesive tape. Space should be allowed for acceleration before the first line, and for slowing down after the second one. When the tape is played, both the time taken to cover the distance, and the number of single steps taken, are measured. It is easiest to take the first foot contact beyond each of the two lines as the start and finish points for both the timing and the counting. This is not strictly accurate from a measurement point of view, but the errors incurred are likely to be trivial. Since the distance, the time and the number of steps are all known, the general gait parameters can be calculated using the formulae:

$$\text{cadence (steps/min)} = \text{steps counted x 60/time (s)}$$
$$\text{stride length (m)} = \text{distance (m) x 2/steps counted}$$
$$\text{velocity (m/s)} = \text{distance (m)/time (s)}$$

Appendix 3 gives a very simple computer program, written in the BASIC language, which permits the simultaneous measurement of all three general gait parameters during a single walk over a measured distance. It may be used for direct observation of the subject, or for measurement when replaying a videotape. The program may need slight modification for some dialects of the BASIC language. A similar program could easily be written in another programming language to suit the computer facilities available.

Timing the gait cycle

A number of systems have been described which perform the automatic measurement of the timing of the gait cycle, sometimes called the temporal gait parameters. Such systems may be divided into two main classes – footswitches and instrumented walkways. Figure 4.3 shows typical data which could be obtained from either type of system.

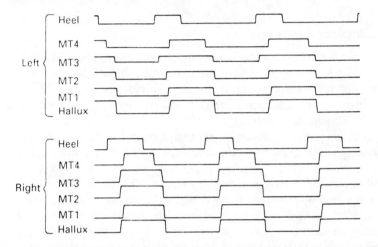

Fig. 4.3 *Output of foot switches under the heel, four metatarsals (MT 1 to MT 4) and hallux of both feet. The switch is on (i.e. the area is in contact with the ground) when the line is high.*

Footswitches

Footswitches are used to record the timing of gait. If one switch is fixed beneath the heel, and one beneath the forefoot, it is possible to measure the timing of heel contact, foot flat, heel off and toe off, and the duration of the stance phase (see Figs 2.2 and 4.3). Data from two or more strides make it possible to calculate cadence and swing phase duration. If switches are mounted on both feet, the single and double support times can also be measured. The footswitches are usually connected through a trailing wire to a microcomputer, although alternatively either a radio transmitter or a portable recording device may be used to collect the data and transfer them to the measuring equipment.

A footswitch is exposed to very high forces, which may cause problems with reliability. This has led to many different designs being tried over the years. A fairly reliable footswitch may be made from two layers of metal mesh, separated by a thin sheet of plastic foam with a hole in it.

When pressure is applied, the sheets of mesh contact each other through the hole, and complete an electrical circuit. Footswitches are most conveniently used with shoes, although suitably designed ones may be taped directly beneath the foot. Small switches may also be mounted in an insole, and worn inside the shoe. In addition to the basic heel and forefoot switches, further switches may be used in other areas of the foot, to give greater detail on the temporal patterns of loading and unloading.

Instrumented walkways

An instrumented walkway is used to measure either the timing of foot contact and/or the position of the foot on the ground. Many different designs have been developed, usually individually built for a single laboratory. The descriptions which follow refer to typical designs, rather than to particular systems.

The conductive walkway is a gait analysis walkway that is covered with an electrically conductive substance such as sheet metal, metal mesh or conductive rubber. Suitably positioned electrical contacts on the subject's shoes complete an electrical circuit. The conductive walkway is thus a slightly different method of implementing footswitches, and provides essentially the same information. Again, the subject usually trails an electrical cable, which connects the foot contacts to a microcomputer. The velocity may be determined independently, by having the body of the subject interrupt the beams of two photoelectric cells, one at each end of the walkway, again connected to the microcomputer. Timing information from the foot contacts is used to calculate the cadence, and the combination of cadence and velocity may be used to calculate the stride length.

A less common arrangement is to have the walkway itself contain a large number of switch contacts, which detect the position of the foot as well as the timing of heel contact and toe off. This has the advantage that no trailing wires are required, and the walkway can be used to measure the two step lengths and the stride length. However, the large number of switch contacts involved may lead to difficulties with reliability. A two-dimensional walkway of this type was built in the author's laboratory, which was able to measure stride width, as well as step length, stride length and gait timing. However, oxidation of the switch contacts made it very unreliable, and it was eventually abandoned.

Two possible sources of inaccuracy when using any form of switch for measuring gait timing are unreliable switch contacts, which may miss events or record them late, and using too slow a microcomputer sampling rate. If the sampling rate is 50 Hz, for example, the switches will only be tested at 20 ms intervals, and the typical 100 ms double support time can only be measured to an accuracy of 20 per cent. It is thus preferable to use a sampling rate in the region of 200–500 Hz. A disadvantage of using either footswitches or conductive walkways is that a trailing wire is usually needed. However, by careful routing of the wire, interference with the gait pattern may be minimized.

Direct motion measurement systems

A number of systems have been described which measure the motion of the body or legs using some form of direct connection to the subject.

The simplest of these systems measures the forward displacement of the trunk by means of a light string which is connected to the back of a belt around the subject's waist. As the individual walks forwards, the string is pulled through an instrument which measures its motion. This may be achieved in a number of ways, including a tacho-generator, which is a form of dynamo, the output voltage of which relates to the instantaneous velocity of the string, or a device such as an optical encoder, which can be used to measure the length of the string as it passes through. Such systems will give the mean velocity of walking, and also the instantaneous velocity as it changes during the gait cycle.

A more elaborate system, based on the same principle, may be used to measure pelvic twist about the vertical axis by having one string attached to each side of the waist. It is also possible to measure lateral displacement in walking by having one or two strings running sideways, although some means must be provided to cope with the forward motion, such as having the subject walk on a treadmill.

Direct connection systems may also be used to measure the motion of the legs. One such system uses perforated paper tape, attached to the heels of both shoes, which is pulled through an optical reader as the subject walks forwards (Law, 1987; Law and Minns, 1989). This gives timing and displacement information for both feet, enabling the calculation of the cadence, the two step lengths, stride length, velocity, stance and swing phase durations, and the two double support times. Another system, which gives similar information, is based on a small trolley which is towed along behind the subject by a loop of cord, the ends of which are attached to the two feet (Klenerman et al., 1988). The cord runs backwards and forwards through a measuring instrument on the trolley, and its movements provide the temporal and distance parameters of gait. The designers of both systems claim that the subject is not aware of any drag produced by the attachments to the feet.

Electrogoniometers

An electrogoniometer is a device for measuring the angle of a joint during walking. The three principle types are based on the potentiometer, the flexible strain gauge, and the resistance of a rubber tube filled with mercury. An associated device, though not strictly an electrogoniometer, is the polarized light goniometer.

The output of an electrogoniometer is usually plotted as a chart of joint angle against time, as shown in Fig. 2.5. However, if measurements have been made from two joints (typically the hip and the knee), the data may be plotted as an *angle/angle diagram*, as shown in Fig. 4.4. This format clarifies the interaction between the two joints, and makes it possible to identify characteristic patterns.

Potentiometer devices

A rotary potentiometer is a variable resistor, of the type used as a radio volume control, in which turning the central spindle produces a change in electrical resistance that can be measured by an external circuit. It can

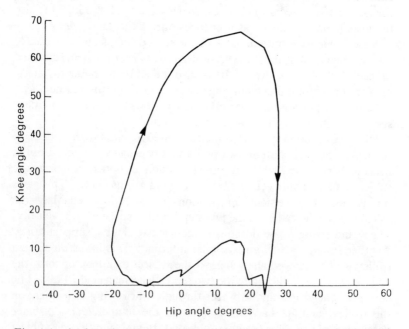

Fig. 4.4   *Angle/angle diagram of the hip angle (horizontal axis) and the knee angle (vertical axis). Normal subject; same data as* Fig. 2.5.

be used to measure the angle of a joint, if it is fixed in such a way that the body of the potentiometer is attached to one limb segment, and the spindle to the other. The electrical output thus depends on the joint position, and the device can be calibrated to measure the joint angle in degrees. Many laboratories have constructed their own electrogoniometers of this type, and there are a few commercially available designs, including those sold by Biokinetics, Chattecx and MIE Medical Research.

Although any joint motion could be measured by an electrogoniometer, they are most commonly used for the knee, and less commonly for the ankle and hip. Fixation is usually achieved by cuffs that wrap around the limb segment above and below the joint. The position of the potentiometer is adjusted to be as close to the joint axis as possible. A single potentiometer will only make measurements in one axis of the joint, but two or three may be mounted in different planes to make multi-axial measurements. Figure 4.5 shows the 'Triax' system, consisting of nine potentiometers in three sets of three, used to measure the three-dimensional motion of the hip, knee and ankle. Trailing wires are normally used to connect the potentiometers to the measuring equipment, which is usually a microcomputer.

Concern has been expressed about the accuracy of measurement provided by these potentiometer devices, since they are subject to a number of possible types of error:

1. The electrogoniometer is fixed to cuffs around the soft tissues, not to the bones, so that the output of the potentiometer does not exactly relate to the true bone-on-bone movement at the joint.
2. Some designs of electrogoniometer will only give a true measurement of joint motion if the potentiometer axis is aligned to the anatomical axis of the joint. This may not be achievable for three reasons:
   a) it may be difficult to identify the joint axis, e.g. because of the depth of the hip joint below the surface
   b) the joint axis may not be fixed, e.g. the 'polycentric' flexion–extension axis of the knee
   c) the rotation axis may be inaccessible, e.g. the internal–external rotation axis of the knee.
3. A joint may, in theory, move with up to six degrees of freedom – that is, it may have angular motion about three mutually perpendicular axes, and linear motion ('translations') in three directions. In practice, the linear motion is usually negligible, particularly at the hip and ankle, and most electrogoniometer systems simply 'lose' any motion which does occur, either through the elasticity of the

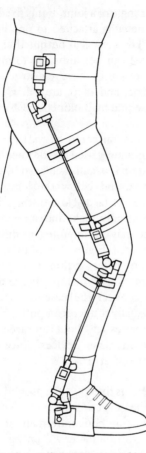

Fig. 4.5 *Subject wearing triaxial goniometers on hip, knee and ankle. Adapted from manufacturer's literature (Chattecx Corporation).*

    mounting cuffs, or through a sliding or 'parallelogram' mechanical linkage. However, where the electrogoniometer axis does not correspond exactly with the anatomical axis, larger linear motions will occur.

4. The output of the device gives a relative angle, rather than an absolute one, and it may be difficult to decide what limb angle should be taken as 'zero', particularly in the presence of a fixed deformity.

Because of these problems, electrogoniometers are more popular in a clinical setting, in which great accuracy is not usually needed, than in the scientific laboratory. Chao (1980) addressed some of these problems, in a defence of the use of potentiometer-based electrogoniometers in gait analysis.

Flexible strain gauges

The flexible strain gauge electrogoniometer (Fig. 4.6), manufactured by Penny and Giles Goniometers Ltd, consists of a flat, thin strip of metal, one end of which is fixed to the limb on each side of the joint being studied. The bending of the metal as the joint moves is measured by strain gauges and their associated electronics. Because of the way in which metal strips respond to bending, the output depends simply on the angle between the two ends, linear motion being ignored. To measure motion in more than one axis, a two-axis goniometer may be used, or two or three separate goniometers may be fixed around the joint, aligned to different planes.

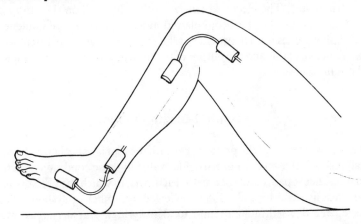

Fig. 4.6   *Subject wearing flexible goniometers on knee and ankle. Adapted from manufacturer's literature (Penny and Giles Goniometers Ltd).*

Mercury-in-rubber devices

A goniometer may be made from an elastic tube, originally made of rubber, but nowadays of plastic. It is filled with mercury and has electrical contacts within its two ends. An external circuit measures the resistance of the mercury column, which increases when the tube is stretched. The main way this device is used as a goniometer is to fix it vertically downwards, in front of the knee, one end being attached to the thigh, and the other to the shin. As the knee flexes and extends, the length of the tube, and hence its resistance, changes. The change in resistance is used to derive the joint angle, using a suitable calibration. One would not expect this type of measurement to be particularly accurate, and there are only a few joint movements for which it is suitable.

Polarized light goniometer

The polarized light goniometer is used to measure the angle between a limb segment and a reference direction, such as the gravitational vertical. By making this measurement for two adjacent limb segments, it is a simple matter to calculate the angle of the joint between them. For example, sensors on the thigh and the shank may be used to measure the knee angle. A photoelectric cell, covered by a sheet of polaroid, is fixed to the limb segment in question. The limb is illuminated by a lamp shining through another sheet of polaroid, which is continuously rotated by an electric motor. The electrical output of the photocell on the limb rises and falls, as the two sheets of polaroid move in and out of alignment. A fixed photocell is used to provide a reference signal, and the difference in phase between this and the output of the limb photocell gives the angle of the limb segment.

Pressure beneath the foot

The measurement of the pressure beneath the foot is a specialized form of gait analysis that may be of particular value in those conditions, such as diabetic neuropathy and rheumatoid arthritis, in which the pressure may be excessive. Lord et al. (1986) reviewed a number of systems which have been devised in an attempt to identify high pressure areas. Most foot pressure measurement systems are floor-mounted. It is more relevant, but also more difficult, to measure the pressure beneath the foot inside the shoe.

The SI unit for pressure is the pascal (Pa), which is a pressure of one newton per square meter. The pascal is inconveniently small, and practical measurements are made in kilopascals (kPa) or megapascals (MPa). Conversions between different units of measurement will be found in Appendix 2.

Lord et al. (1986) pointed out that when making measurements beneath the feet it is important to distinguish between force and pressure (force per unit area). Some of the measurement systems measure the force (or 'load') over a known area, from which the mean pressure over that area can be calculated. However, this mean pressure may be very different from the peak pressure within the area, if high pressure gradients are present, which are often caused by subcutaneous bony prominences such as the metatarsal heads.

A pitfall which must be borne in mind when making pressure measurements beneath the feet is that a subject will normally avoid walking on a painful area. Thus, an area of the foot which had previously experienced a high pressure may, if it becomes painful, show a low pressure when it is tested. However, this will not happen if the sole of the foot is anesthetic, as in diabetic neuropathy. In this condition, very high pressures leading to ulceration may be recorded.

Typical pressures beneath the foot are 80–100 kPa in standing and 200–500 kPa in walking. In diabetic neuropathy pressures as high as 1000–3000 kPa may be found. To put these figures into perspective, the normal systolic blood pressure is below 20 kPa (150 mmHg).

Glass plate examination

Some useful semi-quantitative information on the pressure beneath the foot can be obtained by having the subject stand on, or walk across, a glass plate, which is viewed from below with the aid of a mirror or television camera. It is easy to see which areas of the sole of the foot come into contact with the walking surface, and the blanching of the skin gives an idea of the applied pressure. Inspection of both the inside and the outside of a subject's shoe will also provide useful information about the way the foot is used in walking – it is a good idea to ask patients to wear their oldest shoes when they come for an examination!

Direct pressure mapping systems

A number of low-technology methods of measuring pressure beneath the foot have been described over the years. The Harris or Harris–Beath mat is made of thin rubber, the upper surface of which consists of a pattern of ridges of different heights. Before use, it is coated with printing ink and covered by a sheet of paper, after which the subject is asked to walk across it. The highest ridges compress under relatively light pressures, the lower ones requiring progressively greater pressures, making the transfer of ink to the paper greater in the areas of highest pressure. This gives a semi-quantitative map of the pressure distribution beneath the foot. Two less messy systems are Shutrak (Moore Business Forms), based on a principle similar to a typist's carbon paper, and Fuji Prescale (C. Itoh and Co.), where the subject walks on a pressure sensitive film which is laid on top of a textured mat. A number of related schemes have been used, including one in which the pressure beneath the foot flattens an embossed pattern on a sheet of aluminum foil.

Pedobarograph

The best known floor-mounted system for measuring pressure beneath the foot is the Pedobarograph, which uses an elastic mat, laid on top of an edge-lit glass plate. When the subject walks on the mat, it is compressed onto the glass, which loses its reflectivity, becoming progressively darker with increasing pressure. This darkening provides the means for quantitative measurement. The underside of the glass plate is usually viewed by a television camera, the monochrome image from which is processed to give a display in which different colors correspond to different levels of pressure. A commercial version of the Pedobarograph is sold by Baltimore Therapeutic Equipment.

Load cell systems

A number of systems have been described in which the subject walks across an array of load cells, each of which measures the vertical force

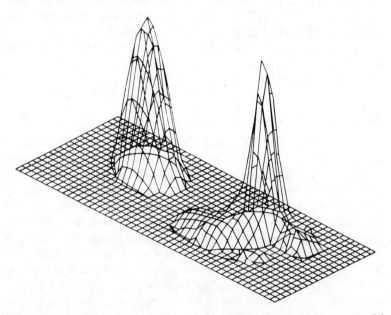

Fig. 4.7   *Pressure beneath a cavus foot on landing from a jump (E.M. Hennig, 5th Biennial Conference, Canadian Society of Biomechanics, 1988).*

beneath a particular area of the foot. Dividing the force by the area of the cell gives the mean pressure beneath the foot in that area. Many different types of load cell have been used, including resistive and capacitative strain gauges, conductive rubber, piezoelectric materials and a photoelastic optical system. Commercially available systems of this type are available from Preston Communications and Novel GmbH. A number of different methods have been used to display the output of such systems, including the attractive presentation shown in Fig. 4.7.

In-shoe devices

The difficult problem of measuring the pressure inside the shoe has been tackled in a number of laboratories, and has led to commercial systems from Electrodynogram Systems and Infotronic. The main difficulties with this type of measurement are the curvature of the surface, a lack of space for the transducers, and the need to run large numbers of wires from inside the shoe to the measuring equipment. For these reasons, most systems of this type measure pressure only in selected areas, in contrast to the floor-mounted systems, which measure it over the whole area beneath the foot.

Electromyography

Electromyography (EMG) is the measurement of the electrical activity of a contracting muscle – the muscle action potential. Since it is a measure of electrical and not mechanical activity, the EMG cannot be used to distinguish between concentric, isometric and eccentric contractions, and the relationship between EMG activity and the force of contraction is far from straightforward. One of the most useful textbooks on the EMG is that by Basmajian (1974).

The three methods of recording the EMG are by means of surface, fine wire and needle electrodes. No attempt will be made to list the manufacturers of EMG equipment, since they are so numerous.

Surface electrodes

Surface electrodes are fixed to the skin over the muscle, the EMG being recorded as the voltage difference between two electrodes. It is usually also necessary to have a grounding electrode elsewhere on the body. Since the muscle action potential reaches the electrodes through the intervening layers of fascia, fat and skin, the voltage of the signal is relatively small, and it is usually amplified close to the electrodes using a very small pre-amplifier. The EMG signal picked up by surface electrodes is the sum of the muscle action potentials from many motor units within the most superficial muscle. Most of the signal comes from within 25 mm of the skin surface, so this type of recording is not suitable for deep muscles such as the iliopsoas. The EMG data may not be very specific, even with superficial muscles, due to interference from adjacent muscles ('crosstalk'). It is safest to regard the signal from surface electrodes as being derived from muscle groups, rather than from individual muscles. There is often a change in the electrical baseline as the subject moves ('movement artifact'), and there may also be electromagnetic interference, for example from nearby electrical equipment.

Fine wire electrodes

Fine wire electrodes are introduced directly into the muscle using a hypodermic needle. Even when the skin is anesthetized with local anesthetic, they can be quite uncomfortable. The wire is insulated except for a few millimeters at the tip. The EMG signal may be recorded in three different ways:

1. Between a pair of wires inserted using a single needle
2. Between two fine wires inserted separately
3. Between a single fine wire and a ground electrode.

The voltage recorded within the muscle is generally higher than that from surface electrodes, particularly if separate wires are used, and there is less interference from movements and electromagnetic fields. The signal is also derived from a fairly small region of a single muscle, generally from a few motor units, a fact which must be taken into account when interpreting the data. Because it is an uncomfortable and invasive technique, fine wire EMG is usually only performed on selected patients.

Needle electrodes

Needle electrodes are generally more appropriate to physiological research than to gait analysis. A hypodermic needle is used, which contains an insulated central conductor. This records the EMG signal from a very local area within the muscle into which it is inserted, normally recording signals from only a single motor unit.

Limitations of EMG

The main problem with the use of any form of EMG is that it is a semi-quantitative technique which gives only an approximation to the strength of contraction of the individual muscles. Many attempts have been made over the years to make it more quantitative, but with only limited success. The other problem with EMG is that it may be quite difficult to obtain satisfactory recordings from a walking subject. This depends partly on the electronic characteristics of the equipment being used, and partly on the skill of the operator in selecting the recording sites and in attaching the electrodes to minimize skin resistance and movement artifacts.

Despite these problems, the information given by EMG can be of considerable value. For example, one form of treatment in cerebral palsy is to transfer the tendon of a muscle to a different position, thereby altering the action of the muscle. When contemplating this type of surgery, it is essential to use EMG first, to make sure that the timing of muscular contraction is appropriate for its new role.

Processing EMG data

For use in gait analysis, it is important that the subject be able to walk naturally while the EMG is being recorded. This may be achieved by transmitting the signals to a receiver ('telemetry'), or by the use of a trailing wire. EMG signals consist of both positive and negative voltage spikes. They may be recorded and displayed in the 'raw' form, or the signal may be processed in some way to produce a relatively smooth trace, the height of which represents the amount of electrical activity present (Fig. 4.8). This processing usually consists of 'envelope detection,' the output of which is sometimes, but erroneously, called the 'integrated EMG.' A better name is 'mean rectified EMG' (MREMG). It consists of the following stages:

1. Filtering with a 'high-pass filter' to remove low frequency signals such as movement artifacts
2. Full-wave rectification, so that the positive and negative spikes do not cancel each other out
3. Filtering with a 'low-pass filter' to remove the spikes, giving a smooth 'envelope' to the signal.

It is important to use appropriate filters for this processing. In particular, the use of too high a cutoff frequency for the first stage filter can lead to

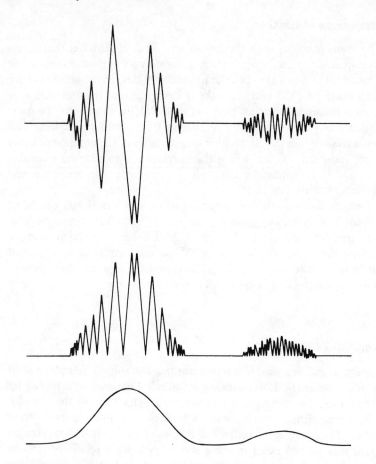

Fig. 4.8   *EMG processing. Top: raw EMG; center: full-wave rectified; bottom: rectified and smoothed.*

the loss of valid EMG data, by suppressing valid low frequency components as though they were 'noise'.

Energy consumption

The most accurate way of measuring the total amount of energy used in performing an activity such as walking is 'whole body calorimetry,' in which the subject is kept in an insulated chamber from which the heat output of the body can be measured. This is, of course, quite impractical, except as a research technique, and the usual way of estimating energy expenditure is based on measuring the body's oxygen consumption. There are also less direct methods, using either mechanical calculations or the measurement of heart rate.

Oxygen consumption

The measurement of oxygen consumption requires an analysis of the subject's exhaled breath. If both the volume of air exhaled and its oxygen content are measured, the amount of oxygen consumed in a given time can be calculated. The amount of carbon dioxide produced can also be measured, and the ratio of the carbon dioxide produced to the oxygen consumed (the 'respiratory quotient') provides information on the type of metabolism which is taking place. Except under abnormal environmental conditions, it is not necessary to measure the oxygen and carbon dioxide contents of the inspired air, since these are almost constant.

The classical method of measuring oxygen consumption and carbon dioxide production is to fit the subject with a noseclip and mouthpiece, and to collect the whole of the expired air in a large plastic or rubberized canvas 'Douglas bag'. If the subject is walking, some means of following them around with the bag is needed. A small sample of the gas in the bag is analyzed, after which the volume of exhalate in the bag is measured. After correcting for temperature, air pressure and humidity, a very accurate estimate of oxygen consumption can be obtained. However, collecting the expirate in this way is uncomfortable for the subject, and the technique is quite unsuitable for some patients. A less cumbersome method, though potentially slightly less accurate, involves again a noseclip and mouthpiece, but uses a portable system that performs continuous gas sampling and volume measurement, so that it is not necessary to collect the whole of the expirate. Either form of gas collection can be achieved using a face mask rather than a mouthpiece and noseclip, although it may be very difficult to prevent leakage, or to detect leakage if it should occur. Occasionally, studies of locomotion are made using a spirometer, in which the subject breathes in and out of an oxygen-filled closed system which absorbs the exhaled carbon dioxide. Since spirometers are not usually portable, they are practical only when using a treadmill.

Mechanical calculations

It was pointed out in Chapter 1 that the expenditure of metabolic energy does not result in the production of an equivalent amount of mechanical work. Indeed, in eccentric muscular activity, metabolic energy is used to absorb, rather than to generate, mechanical energy. Even when muscular contraction is used to do positive work, the efficiency of conversion is relatively low, as well as being variable and difficult to estimate. For these reasons, it is generally unsatisfactory to use mechanical calculations to estimate the total metabolic energy consumed in a complicated activity such as walking. However, this method of estimating energy expenditure is used in some laboratories (Gage et al., 1984), and is known as the 'estimated external work of walking' (EEWW). Calculations of this type are more reliable for activities where the relationship between the muscular contraction and the mechanical output is extremely simple, such as in the concentric contraction of a single muscle.

Even though mechanical calculations are generally unsatisfactory for the estimation of the total energy consumption of the body, the measurement of the energy generation and transfer at individual joints may be of great value in gait analysis. Such measurements may be made using combined kinetic/kinematic systems, which are described at the end of this chapter.

Heart rate monitoring

Rose (1983) stated that heart rate monitoring is a good substitute for the measurement of oxygen uptake, since a number of studies over the years have shown that it is surprisingly accurate. It is certainly much easier to perform, and there are a number of systems available, often based on detecting the pulsatile flow in the capillaries, for example in the finger. Another method of recording heart rate is to monitor the electrocardiogram, using electrodes mounted on the chest. As a general rule, the energy consumption is related to the difference in heart rate between the resting condition and that measured during the exercise. Rather than attempting to relate the change in heart rate directly to energy consumption, some investigators use the 'physiological cost index' (PCI), which is said to be less sensitive to differences between individuals (Steven et al., 1983). It is calculated as follows:

PCI = (heart rate walking – heart rate resting) / velocity

The calculation must be made using consistent units, with the heart rate in beats per minute and the velocity in meters per minute, or the heart rate in beats per second and the velocity in meters per second. The PCI is measured in net beats per meter.

Accelerometers

Accelerometers, as their name suggests, measure acceleration. Typically, they contain a small mass, connected to a stiff spring, with some electrical means of measuring the spring deflection when the mass is accelerated. The type of accelerometer used in gait analysis is usually very small, weighing only a few grams. It normally only measures acceleration in one direction, but they may be grouped together for two- or three-dimensional measurements. Solid-state accelerometers, built as integrated circuits, have recently become available. Because of their small size, they may prove to be of value in gait analysis, and also in providing feedback for future systems involving powered artificial limbs and orthoses.

Typically, accelerometers have been used for gait analysis in one of two ways: to measure transient events, or to measure the motion of the limbs.

Measurement of transients

Accelerometers are very suitable for measuring brief, high acceleration events, such as the heelstrike transient. Johnson (1990) described the development of an accelerometer system to assess the performance of shock-attenuating footwear, which is available as the 'shock meter' from J. P. Biomechanics. The main difficulty with this type of measurement is in obtaining an adequate mechanical linkage between the accelerometer and the skeleton, since slippage occurs in both the skin and the subcutaneous tissues. On a few occasions, experiments have been performed with accelerometers mounted on pins, which were screwed directly into the bones of volunteers. A review of the subject was given by Collins and Whittle (1989).

Measurement of motion

The use of accelerometers for the kinematic analysis of limb motion has been explored by a number of research workers, notably Morris (1973). If the acceleration of a limb segment is known, a single mathematical integration will give its velocity, and a second integration its position, provided both the position and the velocity are known at some point during the measurement period. However, these requirements, combined with the 'drift' from which some accelerometers suffer, has prevented them from coming into widespread use for this purpose. If the limb rotates as well as changes its position, which is normally the case, further accelerometers are needed to measure the angular acceleration.

Gyroscopes

It has been suggested that gyroscopes could be used to measure the orientation of the body segments in space, and that 'rate-gyros' could be used to measure angular velocity and acceleration. The author is not aware of any gait analysis research which has used gyroscopes in this way, but the development of very small solid-state devices may make this an attractive method of measurement in the future.

Force platforms

The force platform, which is also known as a 'forceplate,' is used to measure the ground reaction force as a subject walks across it (Fig. 4.9). Although many specialized types of force platform have been developed, most clinical laboratories use a commercial platform from one of three manufacturers – Kistler Instruments, AMTI or Bertec Corporation. A typical product from one of these firms has a flat rectangular upper surface measuring 400 mm by 600 mm, and the platform is about 100 mm high. To make the upper surface extremely rigid, it is either made of a large piece of metal or of a lightweight honeycomb structure. Within the platform, a number of transducers are used to measure tiny displacements of the upper surface, in all three axes, when force is applied to it. The electrical output of the platform may be provided as either six or eight channels. An eight channel output consists of:

1. Four vertical signals, from transducers near the corners of the platform

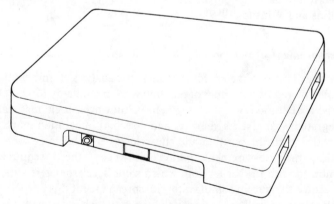

Fig. 4.9 *Force platform. Adapted from manufacturer's literature (AMTI Inc).*

2. Two fore–aft signals from the sides of the platform
3. Two side-to-side signals from the front and back of the platform.

If six channels are provided, they consist of three force vector magnitudes and three moments of force, in a coordinate system based on the center of the platform.

Although it is possible to use the output signal from a force platform directly, for example by displaying it on an oscilloscope, it is much more usual to collect it into a computer, through an analog-to-digital converter. Neither the eight channel nor the six channel force platform output is particularly convenient for biomechanical calculations, and it is usual to convert the data to some other form. In the author's laboratory, the following parameters are calculated and stored for further analysis:

1. Force component along walkway (F_x)
2. Force component across walkway (F_y)
3. Vertical component of force (F_z)
4. Position of center of pressure along walkway (P_x)
5. Position of center of pressure across walkway (P_y)
6. Moment of force in vertical axis about center of pressure (M_z).

Regrettably, no standard has been established for the coordinate systems used for either kinetic or kinematic data, and most laboratories have devised their own.

Ideally, a force platform should be mounted below floor level, so that its upper surface is flush with the floor. If this is not possible, it is usual to build a slightly raised walkway to accommodate the thickness of the platform. It is highly undesirable to have the subject step up onto the platform, and down off it again, since such steps could never be regarded as normal walking. Force platforms are very sensitive to building vibrations, and many early gait laboratories were built in basements to reduce this form of interference. In the author's opinion, this problem has been overemphasized in the past, since although building vibrations can certainly be seen in force platform data, they are negligible when compared with the magnitude of the signals recorded from subjects walking on the platform.

One problem which may be experienced when using force platforms is that of 'aiming.' To obtain good data, the whole of the subject's foot must land on the platform. It is tempting to tell the subject where the platform is, and to ask them to make sure that their footstep lands squarely on it. However, this is likely to lead to a very artificial gait pattern, as the subject 'aims' for the platform. If at all possible, the platform should be disguised so that it is not noticeably different from the rest of the floor, and the subject should not be informed of its presence. This may require a number of walks to be made, with slight adjustments in the starting position, before acceptable data can be obtained.

Where it is required to record from both feet, the relative positioning of two force platforms can be a considerable problem. There is no single arrangement which is satisfactory for all subjects, and some laboratories have designed systems in which one or both platforms can be moved to suit the gait of individual subjects. Figure 4.10 shows the arrangement used in a number of laboratories, which is a reasonable compromise for studies on adults, but is unsatisfactory when the stride length is either very short or very long. For subjects who have a very short stride length, such as children, better results may be obtained if the platforms are mounted with their shorter dimensions in the direction of the walk. Alternatively, the direction of the walk may be altered to cross the platforms diagonally. Despite these strategies, the problem may be insoluble. For example, Gage et al. (1984) observed that 'force plate data

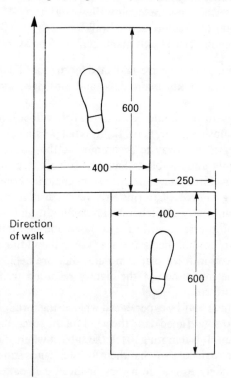

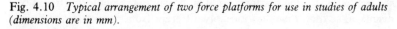

Fig. 4.10 *Typical arrangement of two force platforms for use in studies of adults (dimensions are in mm).*

were discounted because the smaller children frequently stepped on the same platform twice because of their short stride lengths.'

It is often impossible to get the whole of one foot on one force platform, and the whole of the other foot on the other one, without also having unwanted additional steps on one or other platform. To some extent, computer software can be used to 'unscramble' the data when both feet have stepped on one platform, but more commonly it is necessary to use the data from only one foot at a time.

The usual methods of displaying force platform data are:

1. Individual components, plotted against time (see Fig. 2.20)
2. The 'butterfly diagram' (see Fig. 2.6)
3. The center of pressure (see Figs 2.21 and 2.22).

In the latter case, if it is required to superimpose a foot outline in the correct position on the plot, it is necessary also to measure the position of the foot on the force platform, for example by using talcum powder.

A number of things have to be borne in mind when interpreting force platform data. Firstly, although the foot is the only part of the body in contact with the platform, the forces which are transmitted by the foot are derived from the mass and inertia of the whole body, so that the force platform acts as a 'whole body accelerometer.' This means that changes in total body inertia may swamp small changes in ground reaction force due to events occurring within the foot. It may also lead to changes in body inertia being wrongly interpreted as occurring within the foot. For example, fairly high moments are recorded about the vertical axis during the stance phase of gait (Fig. 4.11). These have been erroneously interpreted as being due to local events within the foot, whereas they are mainly derived from the acceleration and retardation of the other leg, as it goes through the swing phase. The reaction to the forces responsible for this acceleration and retardation are transmitted to the floor through the stance phase leg, and appear in the force platform output, principally as a torque about the vertical axis.

When interpreting force platform data, it is also helpful to remember that force is equivalent to the rate of change of momentum. If two objects of identical mass are dropped onto a force platform from the same height, one which bounces will produce a higher ground reaction force than one which does not. This may be surprising, until it is realized that the change in momentum is twice as much for the object which bounces than it is for the other one.

It can be difficult to measure transients, such as the heelstrike, using force platforms. This is because the top plate is usually fairly massive, and does not respond well to very brief forces. The heelstrike transient

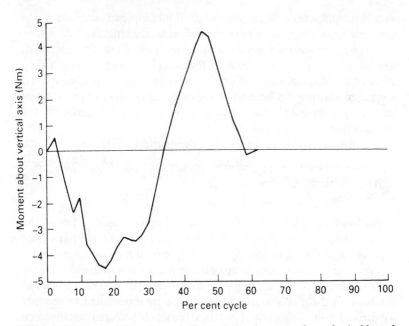

Fig. 4.11 *Moments about the vertical axis during the stance phase of gait. Normal subject, right leg; same data as Fig. 2.20. A positive moment occurs when the foot attempts to move clockwise relative to the floor.*

shown in Fig. 2.8 was undoubtedly attenuated by the limited frequency response of the force platform used to record it. Newer designs are better, as they make use of stiff but lightweight top plates, and they have a higher frequency response. When attempting to measure transients, it is important not to use a low pass filter, which would attenuate transients. If the data are sampled using a computer analog-to-digital convertor, the sampling rate needs to be high enough to record the waveform accurately.

Force platform data by themselves are of limited value in gait analysis. Nevertheless, some laboratories use them empirically, looking for particular patterns in the 'butterfly diagram' (Rose, 1985) or comparing the heights of the different peaks and troughs. Some inferences can also be made from the shapes of the curves of the individual force components. For example, there is an association between stance phase flexion of the knee and a dip at mid stance in the vertical component of force. However, the true value of the force platform is only appreciated when the ground reaction force data are combined with kinematic data.

The combination provides a much more complete mechanical description of the gait than either by itself, and permits the calculation of joint moments and powers, and the use of mathematical modeling to estimate joint forces.

A number of workers have developed devices which are not force platforms, but have the same function. Typically, they consist of a small number of load cells, which are fixed to the sole of the subject's shoe. As the subject walks, the electrical output gives the ground reaction force and the center of pressure. Typically only the vertical component of the ground reaction force is measured, although at least one three-axis system has been described. The advantages claimed are:

1. The ability to measure multiple steps
2. No problems with 'aiming'
3. No risk of stepping on the platform with both feet
4. No risk of missing the platform, either partly or completely.

The disadvantages are the presence of the load cells beneath the feet, and the associated wiring. The coordinate system for these measurements moves with the foot, in contrast to the room-based coordinate system used for kinematic data. This makes it difficult to combine the two types of data, to perform a full biomechanical analysis.

Force platforms may also be used for balance testing and the measurement of postural sway, which are important in some forms of neurological diagnosis. For a complete analysis of the balance mechanism, however, it is necessary to provide some means of moving both the supporting surface and the visual environment.

Table 4.2 gives the accuracy with which a number of gait parameters can be measured, both by a single force platform and by two force platforms.

Kinematic systems

Kinematics is the measurement of movement, or more specifically the geometric description of motion, in terms of displacements, velocities and accelerations. Kinematic systems are used in gait analysis to record the position and orientation of the body segments, the angles of the joints, and the corresponding linear and angular velocities and accelerations.

Following the pioneering work of Marey and Muybridge in the 1870s, photography remained the method of choice for the measurement of

Table 4.2 *Capability and measurement accuracy for different gait parameters by a single force platform and two force platforms.*

Capability	Accuracy
Single force platform	
Components of force (one foot)	High
Center of pressure (one foot)	High
Stance phase duration (one foot)	High
Two force platforms	
Components of force (both feet)	High
Center of pressure (both feet)	High
Cadence	Low
Stride length	Low
Velocity	Low
Stance phase duration (both feet)	High
Swing phase duration (one foot)	High
Swing phase duration (other foot)	Low
Double support time (one side)	High
Double support time (other side)	Low
Step length (one side)	High
Step length (other side)	Low

human movement for about 100 years, until it was displaced by electronic systems. Two basic photographic techniques were used – *cine photography* and *multiple-exposure photography*. As is well known, cine photography is achieved by the use of a series of separate photographs taken in quick succession. Multiple exposure photography has existed in many different forms over the years. It is based on the use of a single photograph, or a strip of film, on which a series of images are superimposed, sometimes with a horizontal displacement between them, to reduce the amount of overlapping. The 1960s and 1970s saw the development of gait analysis systems based on *optoelectronic techniques*, including television, and these have now largely superseded the photographic methods.

The general principles of kinematic measurement are common to all systems, and will be discussed before considering particular systems in detail.

General principles

Kinematic measurement may be made in either two dimensions or three. Three-dimensional measurements normally require the use of two or more cameras, although methods have been devised in which a single camera can be used to make limited three-dimensional measurements.

The simplest kinematic measurements are made using a single camera in an uncalibrated system. Such measurements are fairly inaccurate, but they may be useful for some purposes. Without calibration, it is impossible to measure distances accurately, and such a system is normally used only to measure joint angles in the sagittal plane. The camera is positioned at right angles to the plane of motion, and as far away as possible, to minimize the distortions introduced by perspective. To give a reasonable size image, with a long camera-to-subject distance, a telephoto (long focal length) lens is used. The angles measured from the film are projections of three-dimensional angles onto a two-dimensional plane, and any part of the angulation which occurs out of that plane will be ignored.

A single-camera system can be used to make approximate measurements of distance if some form of calibration object is used, such as a grid of known dimensions behind the subject. Measurement accuracy will be lost by any movement towards or away from the camera, but this effect can again be minimized if the camera is a long distance away from the subject, using a telephoto lens. Angulations of limb segments, either towards or away from the camera, will also interfere with length measurements.

To achieve reasonable accuracy in kinematic measurement it is necessary to use a three-dimensional system, which involves making measurements from more than one viewpoint, and the use of some form of calibration. With the exception of the Coda 3 system (which is internally calibrated), all of the commercial kinematic systems use a three-dimensional calibration object which is viewed by all the cameras. Computer software is used to calculate the relationship between the known three-dimensional positions of the markers on the calibration object and the two-dimensional positions in which they appear in the field of view of the cameras. When a subject walks in front of the cameras, this process is reversed, and the three-dimensional positions can be calculated for those markers which are visible to at least two cameras. When a marker can be seen by only one camera, its three-dimensional position cannot be calculated, although it may be estimated if missing data can be 'interpolated' using data from earlier and later in the walk.

Table 4.3 gives the accuracy with which a number of gait parameters can be measured, both by a single camera kinematic system and by a multi-camera system.

Although there are considerable differences in convenience and accuracy between cine film, videotape, television/computer and optoelectronic systems, the processing of the data is similar, once it has reached the stage of two-dimensional camera coordinates stored in a computer. A number of different calibration schemes have been described, the most widely used being a method known as direct linear transformation (DLT).

All measurement systems, including the kinematic systems to be described, suffer from measurement errors. Measurement accuracy depends to a large extent on the field of view of the cameras, and it also

Table 4.3 *Capability and measurement accuracy for different gait parameters by single camera and multiple camera kinematic systems. Joint angles will be provided for one or both legs, depending on camera configuration.*

Capability	Accuracy
Single camera system	
Cadence	High
Stride length	Medium
Velocity	Medium
Stance phase duration (both feet)	Medium
Swing phase duration (both feet)	Medium
Double support time (both sides)	Low
Step length (both sides)	Medium
Sagittal plane joint angles	Medium
Multi-camera system	
Cadence	High
Stride length	High
Velocity	High
Stance phase duration (both feet)	Medium
Swing phase duration (both feet)	Medium
Double support time (both sides)	Low
Step length (both sides)	Medium
Three-dimensional joint angles	High
Limb segment velocities	High
Limb segment accelerations	Medium
Joint angular velocities	High
Joint angular accelerations	Medium

differs somewhat between the different systems. A typical average value for measurement error might be 2–3 mm in all three dimensions, throughout the whole of a volume large enough to cover a complete gait cycle (Whittle, 1982). Some commercial systems claim to provide much higher accuracy than this, but the author treats such claims with skepticism, particularly when applied to measurements of moving markers under realistic gait laboratory conditions.

Technical descriptions of kinematic systems use, and sometimes misuse, the terms 'resolution,' 'precision' and 'accuracy.' In practical terms, *resolution* means the ability of the system to measure small changes in marker position. *Precision* is a measure of system 'noise,' being based on the amount of variability there is between one frame of data and the next. For the majority of users, the most important parameter is *accuracy*, which describes the relationship between where the markers really are, and where the system says they are!

The commercial systems are sufficiently accurate to measure the positions of the limbs and the angles of the joints. However, the calculation of linear or angular velocity requires the mathematical differentiation of the position data, which magnifies any measurement errors. A second differentiation is required to determine acceleration, and a small amount of measurement 'noise' in the original data leads to wildly erratic and often unusable results for acceleration. The usual way of avoiding this problem is to smooth the position data, using a low-pass filter, before differentiation. This achieves the desired object, but means that any genuinely high accelerations, such as the heelstrike transient, are lost.

Thus kinematic systems are good at measuring position, but poor at determining acceleration, because of the problems of differentiating even slightly noisy data. Conversely, accelerometers are good at measuring acceleration, but poor at estimating position, because of the problems of integrating data with baseline drift. Really accurate data could be obtained by combining the two methods, using each to correct the other, and calculating the velocity from both. However, the author is not aware of any studies which have used this combined approach.

As well as the errors inherent in measuring the positions of the markers, further errors are introduced because considerable movement may take place between a skin marker and the underlying bone. The amount of error this causes in the final result depends on which parameter is being measured. For example, marker movement has little effect on the sagittal plane knee angle, because it causes only a small relative change in the length of the segments, but it may cause considerable errors in transverse plane measurements or in measurements involving shorter segments, such as the foot. Skin movement may also introduce significant errors in the calculation of moments, powers and joint forces. A possibility for the future, as described in Chapter 5, is to correct for marker movement by modeling the movement relative to the underlying bone.

There are two fundamentally different approaches to positioning markers on the limbs. One method is to mount each marker directly on the skin, generally over a bony anatomical landmark, close to the center of rotation of a joint. The position and orientation of the limb segment is approximately defined by a straight line between two markers. The other approach is to fix a set of at least three markers to each limb segment, either directly or mounted on a rigid structure (sometimes called a 'pod'), so that its position and orientation can be determined in three-dimensional space. The movement of one limb segment relative to the next, and the position of the joint center, may then be derived mathematically. The first method is simpler but less accurate, although both suffer from errors due to marker movement. Rigid plates, in particular, may be moved by the contraction of the underlying muscles, and because they are relatively heavy, their inertia causes them to lag behind the limb segment during rapid accelerations. Figure 4.12 shows two possible ways of arranging lower limb markers, one utilizing 'pods' on the thigh and shank, and one based on markers mounted over anatomical landmarks.

Mention has already been made of the use of force platforms for the measurement of postural sway. Kinematic systems may also be used to make this type of measurement in the standing individual.

Photographic systems

The author does not know of any laboratories that still use multiple-exposure photography, but a few still use cine. The major disadvantages of using cine film are the cost of materials, the time taken to process the film, and the effort involved in manually digitizing the data. Its advantages are the relatively low cost of the equipment, the potentially high accuracy obtainable, and the ability to film subjects 'in the field' rather than in the laboratory. For this last reason, cine photography is still popular for measuring sporting activities. Normal cine cameras expose at speeds of up to 25 frames per second, but special high-speed cameras are usually needed for gait analysis, and particularly for sporting activities, to avoid blurring of the image and loss of detail in movements.

The subject is filmed, using one or more cameras, as he or she

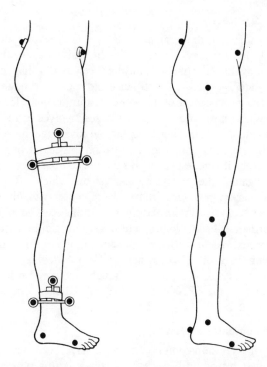

Fig. 4.12 *Typical marker configurations for the pelvis and lower limb. Left: using rigid arrays (Cleveland Clinic). Right: skin-mounted, over anatomical landmarks (University of Oxford).*

performs the activity, whether it be walking or performing some sporting event. If a calibration object is being used, this is also filmed, without altering the camera positions or settings. After the film or films have been processed, the images are digitized by an operator, who views successive frames, identifies the required landmarks, and measures their two-dimensional coordinates. Although this can be done using a ruler, graph paper, or a microscope with a travelling stage, it is most conveniently performed by projecting the film image onto the surface of a 'digitizing tablet' using a specially designed 'analyzing projector.' The operator moves a cursor to the appropriate points in the image, and presses a button to transfer the coordinates of each point into a computer. To avoid inaccuracies caused by frame-to-frame variations in film position, it is common also to digitize the positions of fixed markers within the field of view. A number of different firms supply suitable projectors and digitizers.

Videotape digitizers

The use of videotape to augment visual gait analysis has already been described. Videotape may also be used as the basis for a kinematic system in the same way that cine photography has been used. It has considerable advantages in terms of cost, convenience and speed, although it may not be quite as accurate because of the poorer resolution of a television image compared to cine film, and the further losses involved in recording and replaying a videotape. Another considerable advantage, however, is that it is possible to automate the digitization process using electronic processing of the image, especially if skin markers are used, which show up clearly against the background. Systems which offer automatic tracking of marker positions from videotape are available from Ariel Performance Analysis System, Human Performance Technologies, and Peak Performance Technologies. These systems can be used both as a two-dimensional system with a single camera, or as a three-dimensional system using two or more cameras. Most videotape systems use conventional television equipment, although high-speed systems are also available, such as the SP2000 from Eastman Kodak Company.

Television/computer systems

A number of television/computer systems have been developed since they were first introduced in 1967. Systems of this type are sold by Bioengineering Technology and Systems, HCS Vision Technology, Motion Analysis Corporation, and Oxford Metrics. Although the systems differ in detail, the following description is typical.

Reflective markers are fixed to the subject's limbs, either close to the joint centers, or fixed to the limb segment in such a way as to identify its position and orientation (Fig. 4.12). Close to the lens of each television camera is a light source, which causes the markers to show up as very bright spots. The markers are usually covered in 'Scotchlite,' the material which makes road signs show up brightly when illuminated by car headlights. To avoid the 'smearing' which occurs if the marker is moving, a short exposure time is used. This may be achieved by using one of the following:

1. Stroboscopic illumination
2. A mechanical shutter on the camera
3. A charge-coupled device (CCD) television camera, which is only enabled for a short interval during each frame.

If two or more cameras are used, they are synchronized together. For normal purposes, a conventional television frame rate of 50 Hz or 60 Hz is used, but specialized systems running at higher speeds (typically 200 Hz) are also available. As a general rule, the extra speed is accompanied by a reduction in measurement accuracy.

Each television camera is connected to a special interface board which analyzes each television frame, and locates the edges of any bright spots in the field of view. The position of each marker is taken to be at its 'centroid,' or geometric center (similar in principle to the center of gravity). Either on the interface board itself, or in later processing, the position of the centroid of each marker is calculated using the data from all of the detected edges (Fig. 4.13). Because a large number of edges are used to calculate the position of the centroid, its position can be determined to 'sub-pixel accuracy,' i.e. to a greater accuracy than the resolution of the television image.

The computer stores the marker centroids from each frame of data for each camera, but initially there is no way to associate a particular marker centroid with a particular physical marker, such as that on the right knee. The process of identifying which marker image is which, for each of the cameras, and of following the markers from one frame of data to the next, is known as 'tracking' or 'trajectory following.' The speed and convenience of this process differ considerably from one system to another, and also between software options from a single manufacturer. In the past this has been the least satisfactory aspect of television/computer systems, although software improvements have now made it much more straightforward. The main differences between systems are in whether the tracking process has to be undertaken separately for each camera, or whether the system will perform a 3-D reconstruction of the

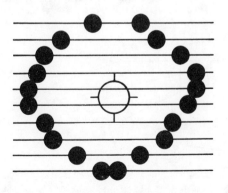

Fig. 4.13   *Location of marker centroid (open circle in center) from the position on successive television lines of the leading and trailing edges of the marker image (solid circles).*

data before the operator is asked to identify which marker is which. The end result of the tracking and 3-D reconstruction process is a computer file of three-dimensional marker positions.

A number of television/computer systems are available which were intended for other purposes, such as aircraft tracking, but which could possibly be adapted for use in gait analysis. Typically, such systems are based on the identification of particular shapes in the video image. They include Videomex from Columbus Instruments, the VRS and Eagle systems from Tau Corporation, and Trackeye from Innovativ Vision AB.

Active marker systems

Another type of kinematic system uses active markers, typically light-emitting diodes (LEDs), and a special optoelectronic camera. Typically, these systems use invisible infra-red radiation, but for the sake of clarity the word 'light' will be used in the following explanation. The camera contains a device which measures the centroid of all the light falling on it. It may be based on either a single two-dimensional camera (Fig. 4.14), or two one-dimensional cameras arranged at right angles.

The LEDs are arranged to flash on and off in sequence, so that only one is illuminated at any instant of time. The cameras are thus able to locate the centroid of each marker in turn, without the need for a 'tracking' procedure to determine which one is which. The penalty for this convenience is the need for the subject to carry a power supply with wires running to each of the markers. Problems may also occur because the camera records not just the light from the LEDs, but any other light

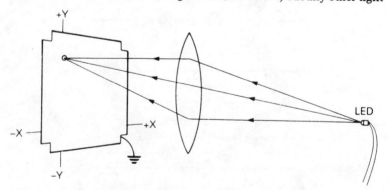

Fig. 4.14   *The image from a light-emitting diode (LED) is focused onto a sensor. Its position in the x and y axes is given by the voltage differences between the appropriate pairs of wires.*

which falls on it. This may include stray ambient light, which is fairly easy to eliminate, and reflections from the markers themselves, which in some systems are difficult both to detect and to eliminate. Despite these problems, active marker systems are usually very accurate, have a high sampling rate, and are convenient to use. Systems of this type are marketed by Log.In, Northern Digital and Selspot.

Optoelectronic scanners

Coda 3, manufactured by Charnwood Dynamics, differs in a number of respects from the other kinematic systems. Its passive reflective markers are very high quality 'corner cube' prismatic reflectors, each of which has a colored filter to enable its identification. The system has three scanners arranged in a fixed orientation (Fig. 4.15). The scanner at each end of the unit measures position in the horizontal axis, and that in the center measures position in the vertical axis. Each scanner projects a rapidly moving plane of light across the subject. The color and timing of the reflections are used to identify and locate the marker positions. Both the

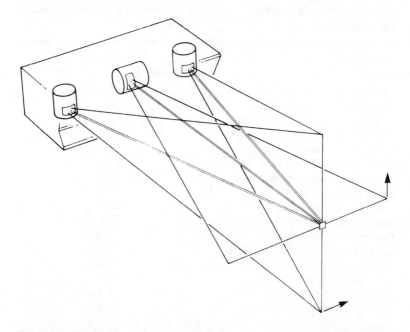

Fig. 4.15 *The Coda 3 system scans the field of view with a vertical fan of light on each side, moving laterally, and a central horizontal fan, moving vertically. From manufacturer's literature (Charnwood Dynamics).*

scanning rate and the accuracy of the system are high, and its output consists of 'real time' 3-D coordinates.

Combined kinetic/kinematic systems

When a kinematic system, such as those described above, is combined with a force platform (which is a kinetic system), the capability of the combined system (Table 4.4) is greater than that of the sum of its component parts (Tables 4.2 and 4.3). The reason for this is that if the relationship is known between the limb segments and the ground reaction force vector, it is possible to perform calculations in which the limb is treated as a mechanical system. While not all commercial systems provide the necessary software to make these calculations, there is the potential to calculate the moments of force and the power generated or absorbed at all the major joints of the lower limb. Such calculations

Table 4.4 *Capability and measurement accuracy for different gait parameters by combined kinetic/kinematic system with multiple cameras and two force platforms.*

Capability	Accuracy
Cadence	High
Stride length	High
Velocity	High
Components of force (both feet)	High
Center of pressure (both feet)	High
Stance phase duration (both feet)	High
Swing phase duration (both feet)	High
Double support time (both sides)	High
Step length (both sides)	High
Sagittal plane joint angles	High
Coronal plane joint angles	High
Limb segment velocities	High
Limb segment accelerations	Medium
Joint angular velocities	High
Joint angular accelerations	Medium
Transverse plane joint angles	High
Joint moments of force	Medium
Joint powers	Medium
Forces in joint structures	Low

require a knowledge of the mass of the limb segments, and the location of their centers of gravity. Direct measurements of these are clearly impossible, but published data, modified to suit the subject's anthropometry, give an acceptable approximation.

The use of mathematical modeling permits estimates to be made of the forces transmitted by and across the various structures of the joints: the tendons, ligaments and articular surfaces. Unfortunately, a large number of unknown factors are involved, especially the internal moment generated by the different muscles, and the extent of any simultaneous contraction by antagonists. For this reason, such calculations can only be approximate, but they can nonetheless be extremely valuable, particularly in clinical and biomechanical research.

In addition to a three-dimensional kinematic system and a pair of force platforms (Fig. 4.16), the best equipped gait analysis laboratories also have facilities for electromyography (EMG). As well as its great value in clinical gait analysis, EMG may be used to refine the output of mathematical models of joint forces. Most models offer a range of possible solutions based on different combinations of active muscles. Even though it is not generally possible to convert EMG signals directly into muscle contraction force, the knowledge that a muscle is either inactive, contracting a little or contracting strongly, makes it possible to eliminate at least some of the possible model solutions, and thereby to improve the reliability of the results.

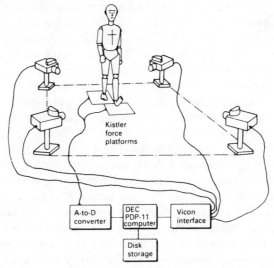

Fig. 4.16   *The Oxford University gait analysis laboratory, with a four-camera Vicon kinematic system and two Kistler force platforms.*

5

Applications of Gait Analysis

The applications of gait analysis are conveniently divided into two main categories: *clinical gait analysis* and *scientific gait analysis*. One problem with this distinction is that it implies that clinical gait analysis is not scientific, and that scientific gait analysis has no clinical value, neither of which is necessarily true. However, there are considerable differences in approach, and even larger differences in equipment, between the clinician who observes gait subjectively, with no technological aids, and the university biomechanics laboratory which uses the techniques of scientific gait analysis as a research tool. Clinical gait analysis has the aim of helping individual patients directly, whereas scientific gait analysis aims to improve our understanding of gait, either as an end in itself, or in order to improve medical diagnosis or treatment in the future.

Even a cursory inspection of the gait analysis literature reveals that significantly more papers have been published on scientific gait analysis than on the clinical applications of the techniques. While it is undoubtedly true that more time and effort has gone into publications relating to scientific gait analysis, the number of published papers in the two areas probably gives a misleading impression, since clinical gait analysis is often less suitable for publication. Clinicians, on the whole, are involved in solving problems for individual patients, which at best can be published only as 'case studies,' whereas scientists are often able to study some aspect of the subject in depth, and to derive publishable conclusions. Another factor which leads to a predominance of research papers is that scientists, far more than clinicians, need to publish their results to maintain their standing in the profession. There is probably more clinical gait analysis being practiced than the literature would suggest, although there is room for much more, and it remains a source of regret that the powerful techniques of gait analysis are only used to benefit relatively small numbers of patients. It has been estimated that, worldwide, not more than one thousand sufferers from cerebral palsy have received any systematic form of gait analysis prior to surgical treatment. However, the medical community is gradually coming to accept that clinical studies of all types need to be supported by objective data. In the fields of orthopedics and rehabilitation, this can often be provided by the techniques of gait analysis.

This chapter is divided into two sections, one on the clinical and the other on the scientific applications of gait analysis. Clinical applications include decision making in a number of patient groups, especially children with cerebral palsy, and the use of gait analysis in the documentation of a patient's condition. Scientific gait analysis is divided into clinical and fundamental research, the former concentrating on disease processes and methods of treatment, the latter on methods of measurement, and the advancement of knowledge in biomechanics, human performance, and physiology.

Clinical gait analysis

Clinical gait analysis involves performing gait analysis on a single person, with the aim of benefiting that person directly. The cornerstone of this process is gait assessment, which seeks to describe, on a particular occasion, the way in which a person walks. This may be all that is required if the aim is simply to document their current status, or it may be just one step in a continuing process, such as the planning of treatment or the monitoring of progress over a period of time.

Davis (1988) pointed out that there are considerable differences between the technical requirements for clinical gait analysis and those for scientific gait analysis. For example, an intrusive measurement system and a cluttered laboratory environment might not worry a fit adult, who was acting as an experimental subject, but could cause significant changes in the gait of a child with cerebral palsy. In scientific gait analysis, it might be acceptable to spend a whole day preparing the subject, making the measurements and processing the data, whereas, in the clinical setting, patients often tire easily and the results are usually needed as quickly as possible. The requirements for accuracy are generally not as great in the clinical setting as they are in the research laboratory, provided the measurement errors are not large enough to cause a misinterpretation of the clinical condition. However, it is essential that those interpreting the data appreciate the likely magnitude of any such errors. Finally, the system must be able to cope with a wide variety of pathological gaits. It is much easier to make measurements on normal subjects than on those whose gait is very abnormal, which may explain why the literature of the subject is dominated by studies of normals! A final and important point is that there is no value in using a complicated and expensive measurement system unless it provides useful information which cannot be obtained in an easier way.

Gait assessment

Rose (1983) made a distinction between gait analysis and gait assessment. He regarded gait analysis as 'data gathering,' and gait assessment as 'the integration of this information with that from other sources for the purposes of clinical decision-making.' This usage of the term 'analysis' differs from that in more technical fields, in which it means 'the processing of data to derive new information.' However, Rose's use of the term is helpful, because it points out that gait assessment is simply one form of clinical assessment. Medical students are taught that clinical assessment is based on three things – history, physical examination and special investigations. In this context, gait analysis is simply a special investigation, the results of which will augment other investigations, such as X-ray reports and blood biochemistry, to provide a full clinical picture. The term 'gait evaluation' is sometimes used instead of gait analysis.

The simplest form of gait assessment is practiced every day in orthopedic and rehabilitation clinics throughout the world. Every time a doctor or therapist watches a patient walk up and down a room, they are performing an assessment of the patient's gait. However, such assessment is often unsystematic, and the most that can be hoped for is to obtain a general impression of how well the patient walks, and perhaps some idea of one or two particular problems. This could be termed an 'informal' gait assessment. To perform a 'formal' gait assessment requires a careful examination of the gait using a systematic approach, if possible augmented by objective measurements. Such gait assessment will usually result in a written report, and the discipline involved in producing such a report is likely to result in a much more carefully conducted analysis.

The gait analysis techniques which are used in clinical gait assessment vary enormously, with the nature of the clinical condition, the skills and facilities available in the individual clinic or laboratory, and the purpose for which the analysis is being conducted. In general, however, gait assessments are made for two possible reasons – for decision making or for documentation.

Clinical decision making

Both Rose (1983) and Gage (1983) suggested that clinical decision making in cases of gait abnormality should involve three clear stages:

1. *Gait assessment:* This starts with a full clinical history, both from the patient and from any others involved, such as a doctor, therapist or

family member. It is followed by a physical examination, with particular emphasis on the musculoskeletal system. Finally, a formal gait analysis is carried out.

2. *Hypothesis formation:* The next stage is the development of a hypothesis regarding the cause or causes of the observed abnormalities. Time needs to be set aside to review the data, and consultation between colleagues, particularly those from different disciplines, is extremely valuable. Indeed, almost all those using gait analysis as a clinical decision-making tool stress the value of the 'team approach.' In forming a hypothesis as to the fundamental problem in a patient with a gait disorder, Rose emphasized that the patient's gait pattern is not entirely the direct result of the pathology, but is the net result of the original problem and the patient's ability to compensate for it. He observed that the worse the underlying problem, the easier it is to form a hypothesis, since the patient is less able to compensate.

3. *Hypothesis testing:* The hypothesis about the cause of the gait abnormality can be tested in two different ways – either by using a different method of measurement, or by attempting in some way to modify the gait. Some laboratories routinely use a fairly complete 'standard protocol,' including videotape, kinematic measurement, force platform measurements, and surface EMG. They will then add other measurements, such as fine wire EMG, where this is necessary to test a hypothesis. Other clinicians start the gait analysis using a simple method, such as videotape, and only add other techniques, such as EMG or the use of a force platform, where they would clearly be helpful. Rose (1983) opposed the use of a standard protocol for all patients, since some of the procedures turn out to have been unnecessary, and there is a risk of ending up with 'an exhausted subject in pain.' The other method of testing a hypothesis is to re-examine the gait after attempting some form of modification, typically by the application of an orthosis, or by paralyzing a muscle using local anesthetic. The ultimate form of gait modification is by surgical operation, with retesting following recovery. However, this is a rather drastic form of 'hypothesis testing,' which can be used only where there is a good reason to suppose that the operation will lead to a definite improvement.

Different types of gait analysis data may be useful for different aspects of the gait assessment. Information on foot timing may be useful to identify asymmetries, and may indicate problems with balance and stability. The general gait parameters give a rough guide to the degree of disability, and may be used to monitor progress or deterioration with the passage of time. The kinematics of limb motion describe abnormal movements, but do not identify the 'guilty' muscles. The most useful measures are probably the joint moments and joint powers, particularly if this information is supplemented by EMG data. Hemiparetic patients may show greater differences between the two sides in muscle power output than in any of the other measurable parameters, including EMG.

Winter (1985) also stressed the need to work backwards from the observed gait abnormalities to the underlying causes in terms of the 'guilty' motor patterns, using both the EMG and the moments about the hip, knee and ankle joints. He offered a method of charting gait abnormalities, and a table giving the common gait disorders, their possible causes, and the type of evidence which would confirm or refute them (Table 5.1). Although the next step, that of treatment, was not considered in detail, he suggested that once an accurate diagnosis had been made, the therapist would be challenged to 'alter or optimize the abnormal motor patterns.'

Table 5.1 *Common gait abnormalities, their possible causes, and evidence required for confirmation (reproduced, with permission, from Winter, 1985).*

Observed abnormality	Possible causes	Biomechanical and neuro-muscular diagnostic evidence
Foot slap at heel contact	Below normal dorsiflexor activity at heel contact	Below normal tibialis anterior EMG or dorsiflexor moment at heel contact
Forefoot or flatfoot initial contact	(a) Hyperactive plantarflexor activity in late swing (b) Structural limitation in ankle range (c) Short step-length	(a) Above normal plantarflexor EMG in late swing (b) Decreased dorsiflexion range of motion (c) See (a), (b), (c) and (d) immediately below
Short step length	(a) Weak push off prior to swing (b) Weak hip flexors at toe off and early swing	(a) Below normal plantarflexor moment or power generation (A2) or EMG during push off (b) Below normal hip flexor moment or power or EMG during late push off and early swing

Observed abnormality	Possible causes	Biomechanical and neuro-muscular diagnostic evidence
	(c) Excessive deceleration of leg in late swing	(c) Above normal hamstring EMG or knee flexor moment or power absorption (K4) late in swing
	(d) Above normal contra-lateral hip extensor activity during contralateral stance	(d) Hyperactivity in EMG of contralateral hip extensors
Stiff-legged weight bearing	Above normal extensor activity at the ankle, knee or hip early in stance*	Above normal EMG activity or moments in hip extensors, knee extensor or plantarflexors early in stance
Stance phase with flexed but rigid knee	Above normal extensor activity in early and mid stance at the ankle and hip, but with reduced knee extensor activity	Above normal EMG activity or moments in hip extensors, and plantarflexors in early and mid stance
Weak push off accompanied by observable pull off	Weak plantarflexor activity at push off. Normal, or above normal, hip flexor activity during late push off and early swing	Below normal plantarflexor EMG, moment or power (A2) during push off. Normal or above normal hip flexor EMG or moment or power during late push off and early swing
Hip hiking in swing (with or without circumduction of lower limb)	(a) Weak hip, knee or ankle dorsiflexor activity during swing (b) Overactive extensor synergy during swing	(a) Below normal tibialis anterior EMG or hip or knee flexors during swing (b) Above normal hip or knee extensor EMG or moment during swing
Trendelenburg gait	(a) Weak hip adductors (b) Overactive hip adductors	(a) Below normal EMG in hip abductors: gluteus medius and minimus, tensor fasciae latae (b) Above normal EMG in hip adductors, adductor longus, magnus and brevis, and gracilis

* Note: There may be below normal extensor forces at one joint but only in presence of abnormally high extensor forces at one or both of the other joints.

Many others working in the field of clinical gait analysis have noted the difficulty of deducing the underlying cause from the observed gait abnormalities, because of the compensations which take place. A number of attempts have been made to simplify this process by using a systematic approach. Computer-based expert systems are very suitable for this type of application, and a number of such systems have already been developed for clinical gait analysis. No doubt the number and quality of such systems will increase in the future.

Cerebral palsy

According to Rose (1983), decision making in cerebral palsy (CP) may be 'significantly affected' by the results of gait analysis, making this condition one of the most important practical applications of the methodology. A very good case for this has been made by Gage (1983), who studied the variable outcome normally seen following the surgical treatment of spastic diplegia, and ascribed it to the fact that 'the spectrum of neurological pathology cannot be differentiated by clinical evaluation alone.'

Gage (1983) also pointed out the desirability of performing all the necessary surgical corrections in a single operative session. Although there are considerable benefits to the patient from having all their operations at once, there is also a risk that it will 'increase the likelihood of a judgement error' (Gage et al., 1984). For this reason, the approach can only be used safely if accurate and complete diagnostic information is available, from a very thorough gait analysis.

Gage et al. (1984) prefer to delay the assessment and treatment of children with spastic diplegia until the gait has matured, generally between the ages of 6 and 8 years. A thorough physical examination is performed, followed by gait analysis, which uses three video cameras, a three-camera television/computer kinematic system, and two force platforms. Surface EMG is recorded, on each side, from the anterior gluteals, gluteus maximus, quadriceps, medial and lateral hamstrings, triceps surae, tibialis anterior and the hip adductors. Where a hip flexion contracture is present, fine wire EMG is also performed on the iliopsoas. Analysis of the results of the gait analysis may indicate the need for fine wire EMG recordings from other muscles (Fig. 5.1).

Having made a detailed diagnosis of the child's functional problems, an appropriate form of treatment is decided, which in many cases involves surgery. Once this has been carried out, and after a suitable recovery period, a further gait assessment is performed. The purpose of this assessment is firstly to determine the success of the treatment, and

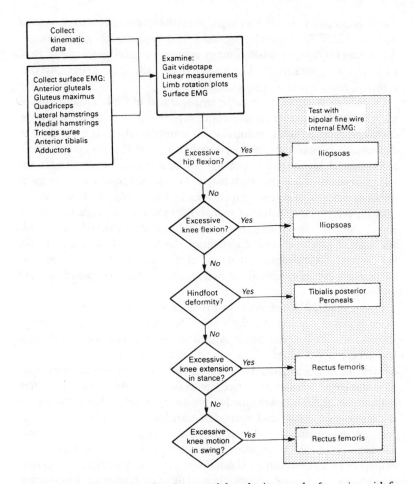

Fig. 5.1 *Gait assessment flowchart, used for selecting muscles for testing with fine wire EMG (Davis, 1988).*

secondly to decide whether the patient would benefit from further surgical procedures, further physical therapy, or the wearing of an orthosis. It also gives the clinical team the opportunity to perform a critical review of the original diagnosis and treatment, to decide whether the correct decisions were made. If an error has been made, it is important to admit it, and to take steps to prevent it from happening again. There is usually no change in EMG postoperatively, and the main criteria for determining the success (or otherwise) of treatment are the general gait parameters and the joint rotations. In subjects in whom it can be measured from the kinematic data, the estimated external work of walking (EEWW) may also be used as a measure of improvement. According to Gage, total body energy consumption is the best measure of the success of treatment. Perry uses three criteria to gauge success: velocity, energy consumption and cosmesis.

Of particular importance in the management of CP is a knowledge of the activity of particular muscles, both to pinpoint 'guilty' muscles and to decide whether a muscle is suitable for transplantation. It is essential to distinguish between the primary pathology and the patient's 'coping responses.' A significant proportion of children undergoing surgery for CP are made worse, rather than better, because inappropriate treatment removes their ability to compensate. For example, a child may be vaulting on one side because of an inability to flex the opposite knee in the swing phase. If the vaulting is erroneously ascribed to a tight heel cord, and treated by a heel cord lengthening, the gait would be made worse, not better. Regrettably, errors of this type are all too common, particularly when treatment is prescribed without the benefit of gait analysis.

Chong et al. (1978) studied the problem of internal rotation of the leg while walking, in children with CP. They demonstrated the value of ambulatory EMG measurements in determining whether or not there was phasic overactivity of the medial hamstrings. As always in CP, a number of different patterns of muscle activity were found in subjects who appeared similar in other respects. They observed that some patients had tightness of the medial hamstrings when walking, but not on clinical examination, and considered that this cause of an internal rotation gait would be 'almost impossible to detect without electro-myography.' They found other patients who had tight hamstrings on clinical examination, but in whom there was no EMG evidence of phasic overactivity. In these patients, they stated that hamstring lengthening would not correct the gait abnormality, and that if the internal rotation was severe enough, a derotation osteotomy of the femur was indicated.

Baumann and Hanggi (1977) showed that gait analysis, using EMG, could distinguish between muscle paralysis and incorrect phasing of muscle contraction. This becomes important when considering trans-planting a muscle tendon to another site to replace a muscle action which is deficient. This can be a very useful method of treatment in CP, but it is vital to know the phasic activity of a muscle before transplanting it. As a general rule, the phasic activity of muscles in CP is fixed, with no possibility of 're-education,' so that if muscle activity is required during the stance phase, the surgery will be 'doomed' if the transposed muscle is only active during swing.

Gage et al. (1987) reported good results from an operation to correct the stiff-legged gait seen in some children with CP. Dynamic EMG is essential to identify those who would benefit from the procedure, which is appropriate only for those in whom there is contraction of the quadriceps throughout the swing phase. The operation involves transfer of the rectus femoris to behind the knee, on either the medial or the lateral side, in connection with lengthening of the hamstrings. According to Gage et al. (1987), this procedure was suggested by Perry.

Winters et al. (1987) described their use of clinical gait assessment to decide the optimal treatment for patients with spastic diplegia, which may result from CP or a number of other conditions. Their use of gait analysis to classify spastic diplegia into four clinical groups, each of which benefits from a different type of treatment, was described in Chapter 3.

Diagnosis of abnormal gait

A number of apparently abnormal gait patterns are, in fact, habits rather than the result of underlying pathology, and the techniques of gait analysis may prove useful in identifying them. Since any pathology affecting the locomotor system generally reduces a person's ability to alter their gait pattern, a variable gait may be suggestive of a habit pattern, and a highly reproducible gait may suggest a pathological process. However, the assessment must take many other factors into account, including the possibility of a pathological process which includes an element of variability, such as ataxia or athetosis.

An example of the use of gait analysis to differentiate between pathological gaits and habit patterns is in the diagnosis of toe walking. Some children walk on their toes, rather than on the whole foot, in a pattern known as 'idiopathic toe walking.' It is important, but also quite difficult, to be able to differentiate between this relatively harmless and self-limiting condition and more serious conditions, such as cerebral palsy. Hicks et al. (1988) stated that earlier attempts to establish the diagnosis, using EMG alone, had not been successful. In their study, they compared the gait kinematics of seven idiopathic toe walkers and seven children with mild spastic diplegia. There was a clear difference between the two groups in the pattern of sagittal plane knee and ankle motion. Both groups had initial contact either by the flat foot or by toe strike, but in the case of the toe walkers, this was due to ankle plantar-flexion, whereas in the children with cerebral palsy it was due to knee flexion. There were also other differences between the two groups, suggesting that gait analysis would be very helpful in making a differential diagnosis.

Gait disorders of the elderly

Cunha (1988) discussed the gait of the elderly, and pointed out that many pathological gait disorders are incorrectly thought to be part of the normal aging process. Identification of an underlying cause, which may be treatable, could result in an improved life for the patient, and a reduced risk of falls and fractures. Cunha classified the causes of the gait disorders of old age as follows: neurological, psychological, orthopedic, endocrinological, general, drugs, senile gait and associated conditions. He described the features of the gait in many conditions affecting the elderly, and suggested a plan for the investigation and management of these patients.

Other conditions

A number of other possible uses for gait analysis in clinical decision making have been suggested over the years, although in most cases the publications dealing with these applications have been largely descriptive, with examples of clinical applications such as the following.

Rose (1983) suggested that gait analysis was useful in the assessment of patients with multiple joint disease. He cited the case of a woman with a stiff, but painless, hip on one side and a painful deformed knee on the other. Gait analysis showed that the stiff hip was causing abnormal loading in the painful knee, suggesting that total knee joint replacement would be 'doomed to early failure.' The recommended course of action was to perform an operation to improve the mobility of the hip before attempting any surgery on the knee. Rose did not, however, see any great value in performing gait analysis on patients requiring the replacement of only a single joint.

Prodromos et al. (1985) investigated the variable clinical results which follow the operation of high tibial osteotomy for osteoarthritis of the knee with a varus deformity. They found that clinical success or failure could be predicted by the preoperative measurement of the coronal plane moment of force at the knee, those patients with a high moment having a significantly worse result, with a recurrence of deformity, than those with a low moment.

Muscle activity may be altered, for example by an increase in the activity of the hamstrings, in patients with a rupture of the anterior cruciate ligament in the knee. EMG of these muscles could be used either as a diagnostic aid, or to monitor the results of ligament replacement surgery.

The management of hemiplegia may be enhanced if a careful assessment of a patient's gait is performed. This may by used as a basis for planning either physical therapy (New York University, 1986), or some form of surgical treatment. Hemiplegic patients tend to walk very inefficiently, so that measuring their joint moments and powers may suggest ways in which training could be used to reduce their energy expenditure. Waters et al. (1979) used fine wire electrodes to measure the EMG activity of all the heads of the quadriceps muscle in hemiplegic patients with a stiff-legged gait. If there was overactivity in only one or two components of the muscle during late stance phase and early swing, good results were obtained by performing appropriate tenotomies. Other neurological conditions could also benefit from the information provided by gait analysis. For example, Rose (1983) stated that gait analysis can make a significant contribution to the management of muscular dystrophy.

Lord et al. (1986) suggested that the measurement of pressure beneath the foot could play a useful part in patient management, by making it possible to provide customized pressure-relieving insoles. The manufacture of these insoles could be achieved either by a craftsman, using information provided by the measurement system, or automatically by some form of computer-aided design and manufacture (CAD–CAM). Having produced the insoles, further pressure measurement could then be used to check their effectiveness. Force platforms have also been used as an aid in the prescription of corrective footwear, by measuring the symmetry of the force patterns between the two feet. Pressure distribution beneath the foot could also be used to monitor the results of treatment for various types of foot disorder, such as a fracture involving the subtalar joint.

A high proportion of longstanding diabetics develop peripheral neuropathy, and many of these go on to develop pressure sores on the feet. Regrettably, these may not be detected until deep-seated infection has occurred, which often leads to amputation and occasionally to death. The detection of these pressure sores does not need gait analysis – it merely needs someone to inspect the feet on a regular basis. However, the detection of high pressures, before ulceration has occurred, can be achieved by a pressure measurement system. Cavanagh et al. (1985) described a method which could be used to detect areas of dangerously high pressure beneath the feet of diabetic patients, prior to the actual formation of an ulcer. Other groups of patients who might benefit from foot pressure monitoring are those with peripheral neuropathy from other causes, such as Hansen's disease, and people with severely deformed feet, for example from rheumatoid arthritis.

The alignment and adjustment of prosthetic limbs may be improved using objective measurements of gait, particularly if repeat trials are performed with different adjustments. Orthotic prescription may be improved if the gait is monitored with the patient wearing different orthoses, for example with different ankle alignments on an ankle–foot orthosis for drop foot (Lehmann et al., 1987). Trials of different orthoses may also form part of the process of hypothesis testing, described above. Kinematic systems may, with care, be used to measure sacroiliac joint movement during walking, for the identification of either increased or decreased motion.

Documentation

Although clinical decision making is the most direct way in which gait analysis may be used to help an individual patient, there are also many instances when simply documenting the current state of a patient's gait may be of value. For some purposes, such documentation may be needed only on a single occasion, where the aim is to quantify a patient's disability. For other purposes, a series of gait assessments over a period of time may be used to monitor either progress or deterioration.

As part of the overall assessment of a patient with a disability, a clinician may require more detail about how well they walk. This type of gait assessment may be directly intended for use in clinical decision making, but it is sometimes performed speculatively, in case it should reveal a treatable cause for the patient's walking disorder.

Gait analysis is frequently used to document the progress of a patient undergoing some form of treatment. The results of the analysis may be used to identify areas where the treatment is ineffective, or it may define an end point for stopping treatment when progress appears to have ceased. Another use of this type of serial analysis is to convince the patient or their relatives that progress has been made, when their own faulty recollection tells them it has not. An objective form of monitoring progress is particularly important for use in the evaluation of novel and controversial methods of treatment, where the enthusiasm of the investigators could lead to errors of judgement!

Gait analysis may form part of the overall documentation of a number of medical conditions, especially those that involve the locomotor system. A deterioration in gait with the passage of time may be detected early, allowing remedial action to be taken. It may also identify clinical signs which should be looked for in other cases of the same condition, particularly if it is very uncommon.

The shape of the ground reaction force curve is sometimes used empirically to monitor progress in rehabilitation. It has been observed that patients walking slowly and painfully have a flat-topped curve, which gradually changes towards the double-peaked normal curve (see Fig. 2.20) as their condition improves.

A regrettable but inescapable part of medical practice today is concerned with litigation. The ability of gait analysis to measure, objectively, at least some aspects of disability, makes it very useful in supporting claims for damages. The author believes he is the first person to have presented gait analysis data as evidence in a court of law in the United Kingdom, in a case in which a child suffered spastic hemiplegia as a result of a motor vehicle accident. Another aspect of litigation concerns medical practice itself, and the risk that a doctor or other health care professional may be sued for negligence. Now that gait analysis has shown its value in the management of cerebral palsy, it may only be a matter of time before it is considered unwise to operate on a case of CP without a preoperative gait assessment.

Scientific gait analysis

The purpose of scientific gait analysis is to improve our understanding of some aspect of gait, either normal or pathological. This includes studying the underlying physiology of the walking process, how it is affected by disease, how walking disorders may be treated, and how best to measure it. Scientific gait analysis may be broadly divided into two disciplines – *clinical research* and *fundamental research* – although there is bound to be some overlap between them.

It is the aim of the remainder of this chapter merely to give a 'flavor' of the types of research which are currently being conducted, the methods they employ, and the ways in which their results may be useful. It is inappropriate, in an introductory book of this type, to attempt an in-depth review of the current state of scientific research in the field, both because much of it would be out of date within months of publication, and also because some of the topics are too obscure for those who are new to the subject. The examples quoted are only intended to be typical of the types of research which are being performed, and not a complete overview. References are given only to publications in readily available journals – at the time of writing, many of these projects have been reported only as oral presentations at scientific meetings.

Clinical research

Clinical research is research involving patients, which is not necessarily expected to benefit those patients directly, but which will hopefully benefit other patients in the future. It is convenient to divide it into the study of diseases and the investigation of methods of treatment.

The study of diseases

A large number of diseases affect the neuromuscular and musculoskeletal systems, and may thus lead to disorders of gait. Among the most important are:

1. Cerebral palsy
2. Parkinsonism
3. Muscular dystrophy
4. Joint diseases of many types
5. Lower limb amputation
6. Stroke
7. Head injury
8. Spinal injury
9. Spina bifida
10. Multiple sclerosis.

Only a small part of the clinical research on these conditions currently involves gait analysis.

When a patient suffers from some condition which affects the locomotor system, their usual response is to make a series of compensations in an attempt to continue walking as well as possible under the circumstances. The difficulties in distinguishing between the original condition and the patient's compensations have already been referred to. The use of gait analysis techniques, particularly the more advanced methods involving the measurement of moments, forces and powers, gives a better insight into the exact nature of the deficit and the way in which the patient is able to compensate for it. This may lead to a better understanding of the pathological condition itself, and may also suggest improved methods of treatment. For example, if a patient is compensating for weakness of one muscle group by using other muscles, physical therapy could be aimed at strengthening those muscles, and perhaps at increasing the patient's skill in using these slightly unnatural movements.

The use of gait analysis in cerebral palsy has already been discussed at length. It is used not only for the management of individual patients, but also in an attempt to improve our understanding of the condition, by dividing it into different diagnostic groups, and by monitoring objectively the results of different forms of treatment. Of particular importance is the production of general hypotheses on the interaction between the neurological deficit, the patient's compensations, and the results of any treatment they might have received (Rose, 1983). An example of this was seen in Chapter 3, where a hypothesis was described which sought to explain the development of crouch gait following Achilles tendon lengthening (see Fig. 3.26). Gait analysis may also be used to classify those patients who have not undergone Achilles tendon lengthening, but who develop a crouch gait as a direct result of muscle imbalance. The classification is made according to the degree of spasticity of the hip flexors and the hamstrings, and aids in the selection of the best form of treatment.

An examination of the moments of force about the joints may also explain some of the gait abnormalities in CP, since deformity may reduce the muscle lever arms, requiring very high muscle forces to be produced in order to produce modest joint moments.

Hemiplegia due to a cerebrovascular accident ('stroke') is a common condition in the elderly, which often severely affects gait. Brandstater et al. (1983) related the temporal parameters of gait, measured using foot-switches, to the clinical severity of the condition, measured on a six point scale. They found that both the walking speed and the symmetry of the gait were strongly correlated with the clinical severity. A number of other measured parameters also correlated with the clinical condition, but these were thought to have changed because of the reduction in walking speed, rather than as a direct result of the stroke. They found, not unexpectedly, that the motor deficit was more important than the sensory deficit in affecting the walking pattern. They also noted that although the relative time spent in single-leg stance on the affected side correlated strongly with the clinical severity, the absolute time spent on that leg did not, because of the slower walking speed in the more severely affected patients. They regarded the measurement of walking speed, in particular, as an 'extremely useful component' of the evaluation of a hemiplegic patient.

Investigation of forms of treatment

For pathological conditions that affect the locomotor system, gait analysis provides an excellent way to compare the results of different forms of treatment, because it is able to provide objective data. Reviewers of scientific publications have started to insist that claims about clinical results should be supported by evidence, making the methods of gait analysis increasingly valuable as a tool for research. There is often a reluctance to use gait analysis for this purpose, since many clinicians associate the term with the use of the complicated and expensive combined kinetic/kinematic systems, of which there are relatively few in the world. However, clinical trials seldom need the detailed data on the functioning of the locomotor system of which these systems are capable. More commonly, simpler methods are perfectly adequate to measure the improvement which takes place after a particular form of treatment. Insufficient use is made of the general gait parameters, yet they are simple to measure and the results are readily amenable to statistical analysis.

One of the commonest ways in which gait analysis is used in clinical research is to perform a direct comparison between two or more methods of treatment. This type of research normally takes the form either of a *prospective study*, which is planned in advance and follows the patients over a period of time, or a *retrospective study*, which examines the consequences of some form of treatment, after the event. As a general rule, prospective studies give more accurate results, since clearly defined procedures can be laid down before the study begins, and the patient's condition is known both before and after treatment. Retrospective studies may suffer from uncertainties as to the pre-treatment condition of the patient, and sometimes as to the exact details of the treatment given. Nevertheless, they may still provide valuable data on the results of a particular form of treatment.

In cerebral palsy, both prospective and retrospective studies may be used to compare different forms of treatment, although retrospective studies may be somewhat more unreliable than usual, because the clinical features of the condition are so variable. A typical way in which gait analysis can be used in cerebral palsy research is to study two groups of patients who undergo different surgical procedures for the same condition. Parameters such as joint angles and general gait parameters may be measured before and after surgery. If the differences between the techniques are small, it would very difficult to demonstrate that one surgical technique was superior without the benefit of objective methods of measurement. The routine follow-up of the results of surgery in cerebral palsy also constitutes clinical research, particularly if the reasons for past successes and failures are analyzed and used as a guide to future treatment.

The comparison of different surgical procedures is also frequently used in studies of joint replacement. A number of studies have shown that, in mechanical terms, the functioning of a prosthetic hip or knee joint tends to be very different from that of a normal joint. Newer designs, and newer methods of surgical implantation, may be tested by the methods of gait analysis to determine how close to 'normal' they are,

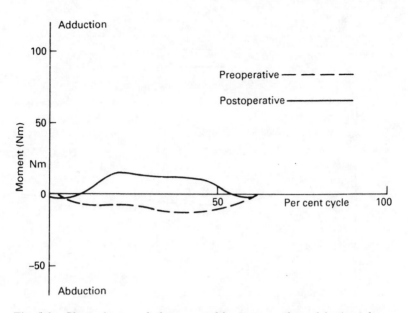

Fig. 5.2 *Change in coronal plane external knee moment from abduction (abnormal) to adduction (normal), following total knee replacement surgery.*

and how much they have improved on earlier designs. The alignment of prostheses plays a large part in determining the stresses to which the prosthesis and the prosthesis–bone interface are subjected. Gait analysis can demonstrate the change in alignment following surgery, as in Fig. 5.2, which shows the change in the knee moment, measured in the coronal plane, before and after a total joint replacement. Gait analysis can also be used to examine different operative techniques. For example, total hip joint replacement may be performed using a number of different approaches, and gait analysis may be used to look for any functional deficits which result, for example, from cutting particular muscles.

As well as its use in comparative trials between two or more forms of treatment, gait analysis is also commonly used simply to quantify the benefit which a patient receives from a particular type of treatment. In such cases, a comparison may be made with pre-treatment data values for that patient, and with comparable results from normal individuals. Figure 5.3 shows the elimination of an excessively high external adduction moment at the knee by the operation of high tibial osteotomy (Jefferson and Whittle, 1989).

The design and prescription of orthose tends to be based much more on art than science, and a significant proportion of orthotic devices are

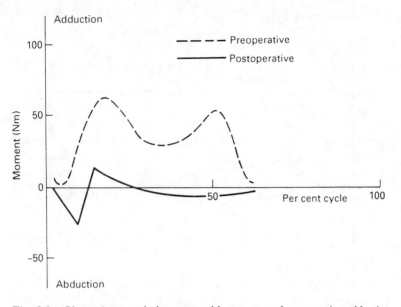

Fig. 5.3 *Change in coronal plane external knee moment from excessive adduction moment to abnormally low biphasic moment, following high tibial osteotomy (Jefferson and Whittle, 1989).*

regrettably not used by the patients for whom they are made. Gait analysis is able to provide an insight into the functioning of orthoses, and to compare different designs. It may also permit improvements to be made to existing designs, such as the Saltiel anterior floor reaction orthosis (Harrington el al., 1984). Some objective studies of orthoses have failed to demonstrate that they have any significant mechanical effects. It then becomes arguable whether the devices really do have no effect, or whether their mode of action is too subtle for the methods of measurement being used.

Gait analysis can be used for the assessment of new or modified forms of lower limb prostheses. It may be used to examine the effects of changing the design of a prosthetic limb, such as by altering the mass distribution (Tashman et al., 1985), or using knee joints with different types of braking mechanism (Hicks et al., 1985). Two types of measurement are of particular value in this type of assessment – the kinematics of motion and the muscle forces the amputee has to produce in order to walk.

Measurement of the power output across the ankle joint is very valuable when comparing different prosthetic foot mechanisms. One of the major differences between the prosthetic foot and the natural foot is the inability of the prosthetic foot to generate power during the push off phase of walking. However, it is able to store energy earlier during the stance phase, and to release it during push off, which may lead to a more natural gait pattern. Gait analysis can be used to examine the energy storage and recovery by prosthetic foot mechanisms, and to determine how well particular types of foot suit different categories of patient. For example, a young subject who has suffered a traumatic amputation is much better able to take advantage of an energy-storing foot mechanism than an elderly subject who has received an amputation for vascular disease.

Electrical stimulation of muscle is becoming increasingly important as a method of treatment in a number of different conditions. It may be used either as a rehabilitation aid, to build up strength in weakened or paralyzed muscles, or as a method of producing useful muscle contraction. The latter is referred to as *functional electrical stimulation* (FES). Electrical stimulation has been used in cerebral palsy to build up strength in paretic muscles, and gait analysis has been used to monitor the resulting improvements. One important use of FES is to act as an 'electronic orthosis' (Muccio et al., 1989), for example for the control of foot-drop by the stimulation of the anterior tibial muscles. The monitoring of the functioning of such a device, and its comparison with mechanical orthoses, would be difficult to do without the use of some form of gait analysis.

Another important use for FES is to enable paralyzed people to walk. The three methods currently under investigation for this purpose are:

1. A purely mechanical approach, using an orthosis
2. The use of FES to stimulate all the necessary muscles
3. A hybrid approach, using an orthosis supplemented by FES.

The categories of paralyzed person most suitable for the restoration of walking by these means are spinal injury and spina bifida, and the best results so far have been obtained in subjects whose paralysis is below about the third thoracic level. Better results have been obtained in children than in adults, because of their lower weight in proportion to their height.

In an investigation of the purely mechanical approach, the author used a combined kinetic/kinematic system to compare two 'walking' orthoses in paraplegic adults (Whittle et al., 1991). There was little difference between the orthoses as far as the general gait parameters were concerned, but the kinetic measurements showed striking differences in the way in which the body and limbs moved, and in the stiffness of the orthosis in the coronal plane. The measurements explained various functional differences between the orthoses, such as the suitability of different walking aids, and also suggested possible directions for further development of the designs.

Optimizing the gait of paralyzed people using some form of FES system involves the use of a complex control system, and gait analysis is required to analyze the success (or otherwise) of the result. One major concern in using FES in paralyzed individuals is that they are usually unable to feel pain, so that it is possible to stimulate the muscles too strongly without the subject being aware of it. This could cause damage to the muscles themselves, or to any of the other structures of the limb, including the bone, which is quite likely to be osteoporotic. The absence of sensation also means that there is no feedback of the position of the limb in space, so that there is a significant risk of the patient falling, if the limb is positioned wrongly.

The use of gait analysis to provide objective outcome measures for different treatments makes it possible not only to compare procedures, but also to test claims for the efficacy of particular forms of treatment. This includes new treatments which require evidence to support (or refute) their efficacy, or established treatments for which there is much anecdotal evidence, but little scientific data to back it up. An example of the former category is the use of selective dorsal rhizotomy, to reduce spasticity in cerebral palsy. An example of the latter is the use of rigid orthotic insoles to treat a wide variety of disorders of the lower limb and spine. Physical therapists use many forms of treatment that appear to be effective, but for which objective scientific evidence is lacking. There is a clear case for investigating the effects of manipulation, serial plastering, exercise training and various forms of electrical and thermal treatment, in a number of common disorders such as athletic injuries and rheumatoid arthritis. Gait analysis is just one of a number of methods which could be used to assess such treatments. In some cases, provision of evidence supporting a treatment may clarify the means by which it produces a beneficial effect, which could in due course lead to ideas for further improvements in treatment.

Gait analysis has also proved to be of value in assessing the effects of drug treatment for diseases affecting the locomotor system. It may be used to titrate the dosage and timing of medication of a single drug, or to compare two drugs. The types of treatment which may be studied in this way include drugs for the treatment of rheumatoid arthritis, which may be studied using joint range of motion, and drugs to reduce spasticity, which may be studied by measuring both the range of motion and the magnitude of the EMG signal. In another application of gait analysis to study drug therapy, Klenerman et al. (1988) described how the time and distance parameters of gait could be used to titrate the dosage of drugs used for the treatment of parkinsonism.

Some of the more specialized gait analysis methods are valuable for assessing footwear. Of particular interest is the pressure distribution beneath the feet, and the way in which it can be modified by suitable insoles. There is also considerable interest, particularly in relation to sporting activities, in the transmission of the heelstrike shock up the leg, and the way in which it may be attenuated by different shoe and insole materials (Johnson, 1988; Johnson, 1990). Since the heelstrike transient is extremely variable from step to step, tests of this type require a large number of measurements to be made, with careful statistical analysis of the results.

Other types of clinical research that may benefit from the use of gait analysis are comparisons between different walking aids, the use of graded exercises for the treatment of osteoporosis (by the estimation of joint forces), and the performance of comparative trials on different postoperative regimes.

Fundamental research

Fundamental research aims to further our knowledge, without having a particular application in mind. However, its long term importance is unarguable, since many of the major advances in medical diagnosis and treatment have been through the application of discoveries made by fundamental research. As related to gait analysis, this type of research is conveniently divided into studies on methods of measurement, and investigations into human biomechanics, performance, and physiology.

Methods of measurement

Although there have been considerable improvements and refinements in gait analysis equipment in the last decade, there have been no completely new methods of measurement. However, much useful fundamental research is being done on the ways in which the data acquired by the present systems can be analyzed. There has been a steady progression during this period from the use of simple measures of walking speed, joint angles and ground reaction force components, through linear and angular accelerations and joint moments of force, to the measurement of joint power output and the forces in the different structures of the limb. Such analyses require the use of mathematical modeling, which at present is usually based on a 'rigid body' analysis.

However, there is a realization that the lower limbs are not rigid, and future models will need to allow for the deformations of the limb structures under load.

The more advanced mathematical models give a number of possible solutions for muscle forces, depending on which combinations of muscles are active. The fact that there is no single solution to the calculations is known as 'indeterminacy,' and is a serious limitation to the use of mathematical modeling in gait. Some of this indeterminacy may be removed by using the EMG signal to determine which muscles are active at different times during the gait cycle. Calculations of power transfers between limb segments may also be refined, using EMG data to determine which muscles are active, and in particular to separate the activity of single-joint and double-joint muscles.

The movement of skin markers relative to the underlying bone is a source of error when using kinematic systems, as was mentioned in Chapter 4. Rather than attempting to prevent the movement, it would be better to accept it as a fact, and to allow for it. It should be possible to model a marker's movements mathematically, knowing its mass, the applied accelerations, the properties of the skin and subcutaneous tissues, and the changes in limb shape as the underlying muscles contract. Once the marker movement relative to the bone has been calculated, the movements of the bone itself could be determined with an improved accuracy.

Refinement of the methods of analysis is not confined to the manipulation of the data a measurement system provides, but may also aim to improve the original data. An example of this, referred to in Chapter 4, is the use of a set of markers on a rigid plate attached to each segment of the limb, rather than having markers stuck directly to the skin over the presumed joint centers. Mathematical techniques can then be used to calculate the apparent centers of the joints, which can be compared with anatomical data on the joint position. Such a rigid plate would be subject to the same types of skin movement as individual markers, but this could again be allowed for in the manner described above.

In order to make the best use of the results of gait analysis, both in clinical management and in clinical research, it is necessary to know the accuracy of the methods being used, and where possible to improve on that accuracy. Error analysis can be performed both theoretically, where an examination is made of the sensitivity of the various measurements to errors in the input data, and experimentally, where the results of an analysis are tested using an alternative method of measurement.

Studies on accuracy and reliability may require the use of methods which can only be done in the research laboratory, such as animal studies, and 'heroic' studies on volunteers. An example of the latter is where pins are inserted into the bones, so that bone motion can be measured without the perturbations of skin movement, and in order to quantify the relationship between the skin markers and the underlying bones. Animal experiments may be used to validate mathematical models, for example in joint force estimation, where the predictions of the model can be compared with direct measurements using implanted transducers.

The verification of methods of measurement may also be performed with the help of volunteer patients, for example when measurements predicted by a mathematical model are compared with the output of transducers mounted in an artificial leg. It is also possible to modify total joint replacement components to contain force-measuring transducers, again giving the opportunity to compare the calculated forces with the results of direct measurements.

Rose (1983) regarded the inability to measure the force of muscular contraction during walking as a 'notable omission' in the methodology of gait analysis. Although this omission has yet to be completely rectified, considerable progress has been made (Perry and Bekey, 1981; Olney and Winter, 1985). A good estimate of muscle force can be made from the EMG signal if sufficient information is available about the current state of the muscle, including its length, its velocity in concentric or eccentric contraction, its fatigue state, and its recent history of contraction. While this amount of information is not readily obtained, it is at least possible in principle, giving hope for increasingly reliable force estimates in the future.

Biomechanics

An important contribution to surgical treatment has been made by the use of gait analysis methods to provide 'design parameters' for orthopedic devices. For example, the design of artificial joints has been greatly aided by estimates of the magnitudes and directions of the forces to which the implanted devices will be subjected. According to Rose (1983), basic research of this type has had a 'seminal effect' on orthopedics, especially in the design of prosthetic joints, replacement ligaments, artificial limbs, and orthotic devices.

Although significant progress has been made in the design and fitting of prosthetic joints, they still have a limited life within the body. One of the most important research applications of gait analysis recently has been to estimate the forces which are transmitted across the interface between the prosthesis and the bone. These forces are thought, at least partly, to be responsible for the mechanical breakdown of the interface, which results in prosthetic loosening. This is a serious problem which often prevents the use of total joint replacement in younger patients. Information on the interface forces can be provided by combined kinetic/kinematic systems, using mathematical modeling. Such information may be used to judge the relative merits of different designs. The hip and knee joints have received the most attention to date, but it is to be hoped that the availability of reliable biomechanical data will eventually lead to improved designs of prosthesis for the first metatarsophalangeal joint, to treat the common condition of hallux valgus.

Ligament replacement surgery has not yet enjoyed the success of joint replacement, but progress continues to be made. The ability to estimate the forces in ligaments makes a significant contribution to the subject, and the mathematical models required for such calculations are constantly being improved. The anterior cruciate ligament of the knee is probably the most important ligament for which this information is needed. Unfortunately, the knee is also one of the most difficult joints to study, because of its complex geometry and the number of two-joint muscles which cross it.

Although a great deal of information now exists on normal walking, further studies continue to be done. Some of these studies simply address details which have received insufficient attention in the past. For example, Hirokawa (1989) measured the variability of the general gait parameters and the effect of applying constraints to cadence, velocity and foot placement. Other research explores new areas or makes measurements in greater detail and with higher accuracy. There has recently been an increasing interest in the joint moments of force, and in the power generation and transfers within the limbs. Modeling and methods of calculation, in becoming more exact, may require additional data, for example by including the metatarsophalangeal joints in power calculations. In analyzing data from patients, it is always necessary to have an adequate baseline of normal data. Since earlier studies were made without the benefit of the new techniques, the normal database becomes deficient, and further studies of normal subjects have to be undertaken.

Research is still needed on the strategies of muscle usage involved in normal walking. Many mathematical models of walking assume some form of optimization, such as minimizing muscle forces or total energy expenditure, but the evidence for the use of such strategies is incomplete. Even when the optimizations used in normal walking are understood, further studies will be needed on the strategies used to compensate for the different pathological conditions which affect gait.

A variety of studies are performed simply to find out how the gait is affected by a particular constraint on the individual, such as fixing a knee in an orthosis or carrying a heavy load. Such research is often conducted as a student project, but once published it may provide a valuable insight into the walking process.

An interesting area of biomechanics research is the study of the transient forces generated when the foot impacts the ground, and the way in which the forces are transmitted up the skeleton. These transients are difficult to measure, using either force platforms or leg-mounted accelerometers, and there is a need for better research on the subject. There are differences between individuals in the magnitude of these transients, due to different strategies in decelerating the swinging foot before initial ground contact. It has been suggested that these transient forces may damage the joints. If this is confirmed, it may be possible to train people to walk in a less damaging way. There has also been considerable interest in the possibility of attenuating such forces by the use of shock absorbing footwear.

Human performance

The study of human performance covers all aspects of human activity, although there is a particular concern with sporting activities. It may include the study of walking under different conditions, for example the changes in joint angle and muscle activity which occur when carrying a load in different ways, walking up and down gradients, and walking on a treadmill.

The study of sporting activities does not normally involve true gait analysis, with the dubious exception of race walking, but it often uses the same equipment and methods of analysis. Furthermore, those involved in gait analysis are often also involved in research in sports biomechanics. With a suitable choice of equipment, it is often possible to use the same kinematic system for gait analysis and for sports biomechanics.

Studies using conventional gait analysis systems have been made of individuals taking part in a wide variety of sports. However, many kinematic systems, which are quite suitable for gait analysis, are neither fast enough nor flexible enough for the measurement of sporting activities. Making kinematic measurements 50 or 60 times per second is generally more than adequate for measurements of walking, but may be much too slow for sporting activities, particularly if it is required to measure velocity and acceleration by the differentiation of position data.

Systems using active markers, and the CODA optical scanning system, have in the past been able to sample at faster rates than the television-based systems. However, some television systems are now available with special cameras running at 200 Hz or more. The limited flexibility of some kinematic systems means that they cannot work in daylight, and they have problems with the constraints on camera positioning which are imposed by the sporting environment. In some sports it is inappropriate, or impossible, to place markers on the athlete. In such cases, the usual solution is to hand-digitize images from film or videotape.

Other gait analysis equipment, such as force platforms and EMG equipment, are usually suitable for the measurement of sporting activities, providing some practical problems can be solved, such as how to keep EMG electrodes in place on sweaty skin!

Physiology research

The methods of gait analysis, particularly the ability to estimate the forces generated by muscles, have proved to be valuable in the study of muscle physiology. The 'holy grail' of this type of research is a reliable method of estimating muscle force from the EMG signal alone, under realistic conditions of muscle contraction.

The power output of muscles during walking, and the power transfers between limb segments, are also of considerable interest. A muscle does external work during concentric contraction, and it absorbs energy during eccentric contraction. During isometric contraction, it does no external work, but by acting as an inelastic strap, a contracting two-joint muscle may be responsible for transferring power from one limb segment to another. The ability to make measurements of power generation, absorption and transfer gives new insights into the ways in which the muscles are used.

Neurophysiology also makes use of the methods of gait analysis, in particular for fundamental studies on the motor control of gait, including the relationships between the hypothetical 'pattern generator' and conscious control. One method of studying a control system is to examine the way in which it responds to perturbations. As related to gait, such perturbations include selective muscle paralysis, limiting the movement of a joint, and the effects of sudden constraints or applied forces. Further understanding of the motor control of gait is gained by studying the development of gait in children (Sutherland et al., 1988), and the walking movements which babies make if held upright with their feet on the ground.

Conclusion

Gait analysis has had a long history, and for much of this time it remained an academic discipline with little practical application. This situation has now changed, and the value of the methodology is now unarguable in certain conditions, especially cerebral palsy. In the future we can hope to see a decrease in the cost and complexity of kinematic systems, and an increasing acceptance by clinicians of the results of gait analysis. The use of these techniques can be expected to increase markedly, both in those conditions for which its value is already recognized, and in a number of other conditions.

Although the present text has focused on gait analysis, this type of measurement equipment may also be used for other purposes – a fact which may be relevant to those trying to obtain funds to set up a gait analysis laboratory! The use of force platforms and kinematic systems for balance and posture testing has already been referred to, as has their use in studying performance in a wide range of sports. Clinical studies have also been made of people standing up, sitting down, ascending and descending stairs. The equipment has been used to measure the movements of the back, not only in walking, but also in the standing and sitting positions. It has also been used to monitor the movements of the upper limbs, both in athetoid patients and in ergonomic studies of reach. Walking is only one of many things which can be done by the musculoskeletal system. It is only sensible to broaden our horizons, and to use the power of the modern measurement systems to study a wide range of other activities.

Appendix 1

Normal ranges for gait parameters

Approximate range (95% limits) for general gait parameters in free-speed walking by normal *female* subjects of different ages

Age (years)	Cadence (steps/min)	Stride length (m)	Velocity (m/s)
13–14	103–150	0.99–1.55	0.90–1.62
15–17	100–144	1.03–1.57	0.92–1.64
18–49	98–138	1.06–1.58	0.94–1.66
50–64	97–137	1.04–1.56	0.91–1.63
65–80	96–136	0.94–1.46	0.80–1.52

Approximate range (95% limits) for general gait parameters in free-speed walking by normal *male* subjects of different ages

Age (years)	Cadence (steps/min)	Stride length (m)	Velocity (m/s)
13–14	100–149	1.06–1.64	0.95–1.67
15–17	96–142	1.15–1.75	1.03–1.75
18–49	91–135	1.25–1.85	1.10–1.82
50–64	82–126	1.22–1.82	0.96–1.68
65–80	81–125	1.11–1.71	0.81–1.61

Approximate range (95% limits) for general gait parameters in free-speed walking by normal *children* (partly based on Sutherland et al., 1988)

Age (years)	Cadence (steps/min)	Stride length (m)	Velocity (m/s)
1	127–223	0.29–0.58	0.32–0.96
1.5	126–212	0.33–0.66	0.39–1.03
2	125–201	0.37–0.73	0.45–1.09
2.5	124–190	0.42–0.81	0.52–1.16
3	123–188	0.46–0.89	0.58–1.22
3.5	122–186	0.50–0.96	0.65–1.29
4	121–184	0.54–1.04	0.67–1.32
5	119–180	0.59–1.10	0.71–1.37
6	117–176	0.64–1.16	0.75–1.43
7	115–172	0.69–1.22	0.80–1.48
8	113–169	0.75–1.30	0.82–1.50
9	111–166	0.82–1.37	0.83–1.53
10	109–162	0.88–1.45	0.85–1.55
11	107–159	0.92–1.49	0.86–1.57
12	105–156	0.96–1.54	0.88–1.60

Appendix 2

Conversions between measurement units

For each parameter, the Système International d'Unités (SI) uses a fundamental unit and a series of multiples and submultiples, in steps of 10^3, as follows:

pico (p) 10^{-12}	kilo (k) 10^3
nano (n) 10^{-9}	mega (M) 10^6
micro (μ) 10^{-6}	giga (G) 10^9
milli (m) 10^{-3}	tera (T) 10^{12}

Length

SI units: millimeter (mm), meter (m), kilometer (km).
Other units: inch (in), foot (ft), mile (mi).

$$1 \text{ mm} = 0.039370 \text{ in}$$
$$1 \text{ m} = 1000 \text{ mm} = 39.370 \text{ in} = 3.2808 \text{ ft}$$
$$1 \text{ km} = 1000 \text{ m} = 0.62137 \text{ mi}$$

Area

SI units: square millimeter (mm^2), square meter (m^2).
Other units: square inch (in^2), square foot (ft^2).

$$1 \text{ mm}^2 = 0.0015500 \text{ in}^2$$
$$1 \text{ m}^2 = 10^6 \text{ mm}^2 = 10.764 \text{ ft}^2$$

Volume

SI units: cubic millimeter (mm^3), cubic meter (m^3).
Other units: milliliter (ml), liter (l), cubic inch (in^3), cubic foot (ft^3), fluid ounce (fl oz), gallon (gal).

$$1 \text{ ml} = 1000 \text{ mm}^3 = 0.061024 \text{ in}^3$$
$$1 \text{ l} = 1000 \text{ ml} = 35.195 \text{ fl oz (UK)} = 33.814 \text{ fl oz (US)}$$
$$1 \text{ m}^3 = 1000 \text{ l} = 219.97 \text{ gal (UK)} = 264.17 \text{ gal (US)} = 35.315 \text{ ft}^3$$

Linear velocity

SI unit: meter per second (m/s or $m \cdot s^{-1}$).
Other units: feet per second (ft/s or $ft \cdot s^{-1}$), kilometers per hour (kph), miles per hour (mph).

$$1 \text{ m/s} = 3.2808 \text{ ft/s} = 3.6000 \text{ kph} = 2.2369 \text{ mph}$$

Linear acceleration

SI unit: meter per second per second (m/s^2 or $m \cdot s^{-2}$)
Other units: feet per second per second (ft/s^2 or $ft \cdot s^{-2}$)

$$1 \text{ m/s}^2 = 3.2808 \text{ ft/s}^2$$

Acceleration due to gravity

$$g = 9.80665 \text{ m/s}^2 = 32.174 \text{ ft/s}^2$$

Mass

SI units: gram (g), kilogram (kg).
Other units: ounce (oz), pound (lb), slug.

$$1 \text{ g} = 0.035274 \text{ oz}$$
$$1 \text{ kg} = 1000 \text{ g} = 2.2046 \text{ lb} = 0.068522 \text{ slug}$$

Force

SI unit: newton (N).
Other units: dyne (dyn), kilogram force (kgf)*, − pound force (lbf), poundal (pdl).

$$1 \text{ N} = 10^5 \text{ dyn} = 0.10197 \text{ kgf} = 0.22481 \text{ lbf} = 7.2330 \text{ pdl}$$

Pressure

SI units: pascal (Pa), kilopascal (kPa), megapascal (MPa).
Other units: bar, millimeter of mercury (mmHg), inch of water (in. H_2O), physical atmosphere (atm), pound force per square inch (psi).

$$1 \text{ Pa} = 1 \text{ N/m}^2 = 10^{-5} \text{ bar}$$
$$1 \text{ kPa} = 1000 \text{ Pa} = 4.0146 \text{ in.} H_2O = 7.5006 \text{ mmHg}$$
$$1 \text{ MPa} = 1000 \text{ kPa} = 9.8692 \text{ atm} = 145.04 \text{ psi}$$

* The kilopond (kp) is identical to the kilogram force.

Energy

SI units: joule (J), kilojoule (kJ), megajoule (MJ).
Other units: erg, calorie, kilocalorie (kcal)*, British thermal unit (Btu), kilowatt hour (kWh).

$$1 \text{ J} = 10^7 \text{ erg} = 0.23892 \text{ cal}$$
$$1 \text{ kJ} = 1000 \text{ J} = 0.23892 \text{ kcal} = 0.94781 \text{ Btu}$$
$$1 \text{ MJ} = 1000 \text{ kJ} = 0.27778 \text{ kWh}$$

Power

SI units: watt (W), kilowatt (kW).
Other unit: horsepower (hp).

$$1 \text{ W} = 1 \text{ J/s}$$
$$1 \text{ kW} = 1.3410 \text{ hp}$$

Plane angle

SI unit: radian (rad)
Other unit: degree (°)

$$180° = \pi \text{ rad}$$
$$1 \text{ rad} = 180/\pi = 57.296°$$
$$1° = 0.017453 \text{ rad}$$

---

* Identical to Calorie, used in dietary calculations.

Appendix 3

Computer program for general gait parameters

The program below is written in BASIC, and should work on most computers. To achieve this, it has been necessary to use unfashionable programming techniques such as GOTO statements – apologies! Two things may need to be changed under some versions of BASIC – the methods of measuring time (TIMER) and of detecting a key being pressed (INKEY$). If the distance over which the measurements are made is always the same, it can be incorporated into the program, rather than having to be entered each time.

The program is started by typing RUN, and the distance is entered before the patient begins to walk. They should reach their normal walking speed *before* crossing the first timing mark. As the first foot contacts the ground beyond this mark, a letter key is pressed (any one will do). Then the space bar is pressed for *every* foot contact until the second timing mark is passed by one of the feet. As this foot contacts the ground, a letter key, rather than the space bar, is pressed. The program then gives the cadence, stride length and velocity. This method is not completely accurate, because of differences in the relative position of the foot and the timing marks at the two ends, but it should be adequate for most purposes.

```
10 PRINT "WALK: A PROGRAM TO MEASURE THE
   GENERAL GAIT PARAMETERS"

20 PRINT

30 PRINT "DISTANCE IN METERS BETWEEN START AND
   FINISH MARKERS";

40 INPUT DIST

50 PRINT

60 PRINT "PRESS ANY LETTER KEY TO START TIMING"

70 PRINT
```

```
80 A$ = INKEY$

90 IF A$ < "A" THEN GOTO 80

100 START = TIMER

110 NSTEPS = 0

120 PRINT "PRESS SPACE BAR FOR EACH STEP – ANY
    LETTER KEY TO FINISH"

130 PRINT

140 A$ = INKEY$

150 IF (A$ < > " ") AND (A$ < "A") THEN GOTO 140

160 NSTEPS = NSTEPS + 1

170 IF A$ = " " THEN GOTO 140

180 DURATION = TIMER – START

190 PRINT "CADENCE: "; INT(60 * NSTEPS / DURATION +
    .5); " STEPS/MIN"

200 PRINT "STRIDE LENGTH: "; INT(200 * DIST / NSTEPS
    + .5) / 100; "METERS"

210 PRINT "VELOCITY: "; INT(100 * DIST / DURATION +
    .5) / 100; " METERS/SEC"

220 END
```

Appendix 4

Addresses of suppliers

A list of this type is bound to be out of date even by the time it is published. However, it is offered in the hope that it will provide at least a starting point for locating firms which sell gait analysis equipment. The list does not include suppliers of EMG equipment or analytical software, except where these form part of a kinematic system.

Telephone numbers are given in the agreed international format, where '+' proceeds the country code, which is then followed by the area code and local number.

Advanced Mechanical Technology, Inc.
(AMTI force platforms)
151 California Street
Newton
Massachusetts 02158
USA
Telephone: +1–617–964–2042

Ariel Performance Analysis System, Inc.
(Videotape-based kinematic system)
6 Alicante
Trabuco Canyon
California 92679
USA
Telephone: +1–714–858–4216

Baltimore Therapeutic Equipment Co.
(Pedobarograph: foot pressure measurement system)
7455 L New Ridge Road
Hanover
Maryland 21076
USA
Telephone: +1–301–850–0333

Bertec Corporation
(Bertec force platforms)
819 Loch Lomond Lane
Worthington
Ohio 43085
USA
Telephone: +1–614–436–9966

Bioengineering Technology and Systems
(Elite: television-based kinematic system)
Via Capecelatro 66
20148 Milano
Italy
Telephone: +39–2–4047896

Biokinetics Inc.
(MIE potentiometer-based electrogoniometer)
1710 Westminster Way
Annapolis
Maryland 21401
USA
Telephone: +1–301–849–3401

Charnwood Dynamics Ltd.
(Coda 3: kinematic system using optical scanners)
63 Forest Road
Loughborough
Leicestershire LE11 3NW
England
Telephone: +44–509–233224

Chattecx Corporation
(Triax: electrogoniometer)
101 Memorial Drive
PO Box 4287
Chattanooga
Tennessee 37405
USA
Telephone: +1–615–870–2281

Columbus Instruments
(Videomex: television/computer tracking system)
PO Box 44049
Columbus
Ohio 43204
USA
Telephone: +1–614–488–6176

Eastman Kodak Company
(SP2000: high-speed videotape system)
11633 Sorrento Valley Road
San Diego
California 92121
USA
Telephone: +1–619–535–2909

Electrodynogram Systems, Inc.
(Electrodynogram: in-shoe pressure measurement system)
1011 Grand Boulevard
Deer Park
New York 11729
USA
Telephone: +1–516–667–1200

HCS Vision Technology BV
(PRIMAS: television-based kinematic system)
Hurksestraat 18d
5652 AK Eindhoven
The Netherlands
Telephone: +31–40–521637

Human Performance Technologies, Inc.
(In-shoe pressure measurement system and videotape–based kinematic
system)
12 Technology Drive
Suite 6
East Setauket
New York 11733
USA
Telephone: +1–516–689–6521

Infotronic
(Computer Dyno Graphy: in-shoe pressure measurement system)
PO Box 73
7650 AA Tubbergen
The Netherlands
Telephone: +31–53–893478

Innovativ Vision AB
(TrackEye: image analysis system)
Teknikringen 1
S–583 30 Linkoping
Sweden
Telephone: +46–13–214060

C. Itoh & Co. (America) Inc.
(Fuji Prescale: pressure-sensitive film)
335 Madison Avenue
New York
New York 10017
USA
Telephone: +1–212–818–8000

J.P. Biomechanics
(Shock meter)
J.C. Peacock & Son Ltd.
Clavering Place
Newcastle upon Tyne
England
Telephone: +44–91–2329917

Kistler Instrumente AG
(Kistler force platforms)
CH–8408 Winterthur
Switzerland
Telephone: +41–52–831111

Kistler Instrument Corp.
(Kistler force platforms)
75 John Glenn Drive
Amherst
New York 14120
USA
Telephone: +1–716–691–5100

Log.In srl
(CoStel: kinematic system using active markers)
Via Aurelia 714
00165 Roma
Italy
Telephone: +39–6–6807044

MIE Medical Research Ltd.
(MIE potentiometer-based electrogoniometer)
6 Wortley Moor Road
Leeds LS12 4JF
England
Telephone: +44–532–793710

Moore Business Forms, Inc.
(Shutrak: pressure-sensitive paper)
1205 N Milwaukee Ave
Glenview
Illinois
USA

Motion Analysis Corporation
(ExpertVision: television-based kinematic system)
3650 North Laughlin Road
Santa Rosa
California 95403
USA
Telephone: +1–707–579–6511

Northern Digital Inc.
(Watsmart and Optotrak: kinematic systems with active markers)
403 Albert Street
Waterloo
Ontario N2L 3V2
Canada
Telephone: +1–519–884–5142

Novel GmbH
(EMED: foot pressure measurement system)
Beichstr, 8
8000 München 40
West Germany
Telephone: +49–89–390102

Oxford Metrics
(Vicon: television-based kinematic system)
Unit 8
7 West Way
Oxford OX2 0JB
England
Telephone: +44–865–244656

Oxford Metrics Inc.
(Vicon: television-based kinematic system)
14206 Carlson Circle
Tampa
Florida 33626
USA
Telephone: +1–813–855–2910

Peak Performance Technologies Inc.
(Peak Performance: videotape-based kinematic system)
7388 S Revere Parkway
Suite 801
Englewood
Colorado 80112
USA
Telephone: +1–303–799–8686

Penny and Giles Goniometers Ltd.
(Flexible strain-gauge electrogoniometer)
Newbridge Road Industrial Estate
Blackwood
Gwent NP2 2YD
United Kingdom
Telephone: +44–495–228000

Preston Communications Ltd.
(Musgrave Footprint: foot pressure measurement system)
New Ross
Dinbren Road
Llangollen
Clwyd LL20 9TF
United Kingdom
Telephone: +44–978–861480

Selspot AB
(Selspot: kinematic system using active markers)
Sallarangsgatan 3
S–431 37 Molndal
Sweden
Telephone: +46–31–870710

Selspot Systems Ltd.
(Selspot: kinematic system using active markers)
21654 Melrose
Southfield
Michigan 48075
USA
Telephone: +1–313–355–5900

Tau Corporation
(VRS and Eagle: television/computer image analysis systems)
485 Alberto Way
Los Gatos
California 95032
USA
Telephone: +1–408–395–9191

Glossary of medical terms

Ankylosis – destruction of a joint, either by surgery or disease, resulting in a total loss of movement.

Ataxia – loss of coordination, due to disease of the central nervous system.

Atherosclerosis – partial or complete blockage of arteries by the fatty substance atheroma.

Athetosis – disorder of the central nervous system, resulting in continual uncoordinated movements of the limbs.

Calliper – orthosis (q.v.) fitted around one or both legs to compensate for weakness or paralysis; also spelled 'caliper.'

Cerebellar ataxia – ataxia (q.v.) due to disease of the cerebellum.

Cerebral palsy – neurological disorder with spasticity (q.v.) and incoordination, caused by brain damage around the time of birth.

Cerebrovascular accident – brain damage due to blood clot or hemorrhage; also known as 'stroke.'

Congenital – present at birth.

Congenital dislocation of hip – congenital condition in which the head of femur is not properly contained in the acetabulum.

Contracture – reduction in the range of motion at a joint, due to restriction by inelastic connective tissue.

Coxa vara – abnormal angulation of the upper end of the femur, making the femoral neck too horizontal.

Derotation osteotomy – operation in which a long bone is cut through and held in a partially rotated position while it heals.

Diabetes – disease in which control of blood sugar is imperfect, through lack of, or abnormal response to, the hormone insulin.

Diabetic neuropathy – loss of function of peripheral nerves, especially sensory nerves in the feet, due to diabetes.

Diplegia – paralysis of both arms or both legs.

Hallux valgus – deformity affecting the great toe, in which the metatarsophalangeal joint is in valgus; also known as 'bunion.'

Hansen's disease – bacterial infection causing progressive destruction of nerves and sensory loss; also known as 'leprosy.'

Hemiparesis – partial paralysis affecting one side of the body.

Hemiplegia – complete paralysis affecting one side of the body, usually due to brain disease or injury.

High tibial osteotomy – operation to change the alignment of the upper tibia, to alter the direction of forces acting on the knee.

Intention tremor – involuntary tremor affecting the performance of fine movements, especially by the hands.

Lordosis – forward curvature of the spine, causing the normal concavity of the lumbar region of the back.

Metatarsalgia – severe pain beneath the metatarsal heads, typically caused by nerve compression.

Multiple sclerosis – progressive neurological disease, caused by a patchy loss of myelin from neurons in brain and spinal cord.

Muscular dystrophy – progressive wasting disease affecting muscles, usually hereditary.

Orthosis – an external support for some part of the body; also known as 'brace.'

Osteoarthritis – degenerative disease affecting the joints, with pain, stiffness and deformity.

Osteotomy – surgical operation involving cutting through a bone.

Paraplegia – complete paralysis of the legs, usually due to disease or injury of the spinal cord.

Paresis – partial paralysis.

Parkinsonism – progressive degenerative neurological condition, involving tremor and weakness; also known as 'Parkinson's disease.'

Peripheral neuropathy – loss of function due to degeneration of the nerves of the hands and feet.

Poliomyelitis – virus disease causing destruction of lower motor neurons, with muscle atrophy and paralysis.

Pressure sore – ulcer (q.v.) forming on an area of skin exposed to sufficient pressure to cut off the blood supply.

Prosthetic joint – an artificial joint, implanted in place of the natural joint.

Prosthetic limb – an artificial limb, usually fixed onto a stump following amputation.

Quadriplegia – paralysis of both arms and both legs; also known as 'tetraplegia.'

Rheumatoid arthritis – a connective tissue disease causing painful inflammation of the joints.

Selective dorsal rhizotomy – operation to reduce spasticity, by cutting sensory nerve roots involved in the stretch reflex.

Slipped femoral epiphysis – hip disease involving displacement of the articular surface of the femoral head, typically in teenagers.

Spasticity – involuntary resistance of muscles to being stretched.

Spina bifida – incomplete development of the vertebrae, resulting in damage to the associated spinal cord.

Stroke – see 'cerebrovascular accident.'

Subluxation – partial dislocation of a joint, the joint surfaces being able to move in and out of the correct alignment.

Tabes dorsalis – loss of sensation and proprioception, due to effect of syphilis on the sensory pathways of the spinal cord.

Tenotomy – surgical cutting of a tendon, either to free it completely, or to lengthen or shorten it.

Tetraparesis – partial paralysis affecting all four limbs.

Tetraplegia – see 'quadriplegia.'

Ulceration – loss of covering skin or epithelium.

References

Basmajian J. V. (1974). *Muscles Alive: Their Functions Revealed by Electromyography*. Baltimore: Williams and Wilkins.

Baumann J. U., Hanggi A. (1977). A method of gait analysis for daily orthopaedic practice. *Journal of Medical Engineering and Technology*, **1**, 86–91.

Brandstater M. E., de Bruin H., Gowland C., Clark B. M. (1983). Hemiplegic gait: analysis of temporal variables. *Archives of Physical Medicine and Rehabilitation*, **64**, 583–587.

Breakey J. (1976). Gait of unilateral below-knee amputees. *Orthotics and Prosthetics*, **30**, 17–24.

Bresler B., Frankel J. P. (1950). The forces and moments in the leg during level walking. *American Society of Mechanical Engineers Transactions*, **72**, 27–36.

Cavagna G. A., Margaria R. (1966). Mechanics of walking. *Journal of Applied Physiology*, **21**, 271–278.

Cavanagh P. R., Hennig E. M., Rodgers M. M., Sanderson D. J. (1985). The measurement of pressure distribution on the plantar surface of diabetic feet. In *Biomechanical Measurement in Orthopaedic Practice* (Whittle M., Harris D., eds.) Oxford, Clarendon Press, pp. 159–166.

Chao E. Y. S. (1980). Justification of triaxial goniometer for the measurement of joint rotation. *Journal of Biomechanics*, **13**, 989–1006.

Chong K. C., Vojnic C. D., Quanbury A. O., et al. (1978). The assessment of the internal rotation gait in cerebral palsy. *Clinical Orthopaedics and Related Research*, **132**, 145–150.

Collins J. J., Whittle M. W. (1989). Impulsive forces during walking and their clinical implications. *Clinical Biomechanics*, **4**, 179–187.

Cunha U. V. (1988). Differential diagnosis of gait disorders in the elderly. *Geriatrics*, **43**, 33–42.

Davis R. B. (1988). Clinical gait analysis. *IEEE Engineering in Medicine and Biology Magazine*. September, 35–40.

Gage J. R. (1983) Gait analysis for decision-making in cerebral palsy. *Bulletin of the Hospital for Joint Diseases Orthopaedic Institute*, **43**, 147–163.

Gage J. R., Fabian D., Hicks R., et al. (1984). Pre- and postoperative gait analysis in patients with spastic diplegia: a preliminary report. *Journal of Pediatric Orthopedics*, **4**, 715–725.

Gage J. R., Perry J., Hicks R. R., et al. (1987). Rectus femoris transfer to improve knee function of children with cerebral palsy. *Developmental Medicine and Child Neurology*, **29**, 159–166.

Garrison F. H. (1929). *An Introduction to the History of Medicine*. Philadelphia: Saunders.

Harrington E. D., Lin R. S., Gage J. R. (1984). Use of the anterior floor reaction orthosis in patients with cerebral palsy. *Orthotics and Prosthetics*, **37**, 34–42.

Hicks R., Tashman S., Cary J. M., et al. (1985). Swing phase control with knee friction in juvenile amputees. *Journal of Orthopaedic Research*, **3**, 198–201.

Hicks R., Durinick N., Gage J. R. (1988). Differentiation of idiopathic toe–walking and cerebral palsy. *Journal of Pediatric Orthopedics*, **8**, 160–163.

Hirokawa S. (1989). Normal gait characteristics under temporal and distance constraints. *Journal of Biomedical Engineering*, **11**, 449–456.

Inman V. T., Ralston H. J., Todd F. (1981). *Human Walking*. Baltimore: Williams and Wilkins.

Jefferson R. J., Whittle M. W. (1989). Biomechanical assessment of unicompartmental knee arthroplasty, total condylar arthroplasty and tibial osteotomy. *Clinical Biomechanics*, **4**, 232–242.

Johnson G. R. (1988). The effectiveness of shock–absorbing insoles during normal walking. *Prosthetics and Orthotics International*, **12**, 91–95.

Johnson G. R. (1990). Measurement of shock acceleration during walking and running using the shock meter. *Clinical Biomechanics*, **5**, 47–50.

Klenerman L., Dobbs R. J., Weller C., et al. (1988). Bringing gait analysis out of the laboratory and into the clinic. *Age and Ageing*, **17**, 397–400.

Krebs D. E., Edelstein J. E., Fishman S. (1985). Reliability of observational kinematic gait analysis. *Physical Therapy*, **65**, 1027–1033.

Law H. T. (1987). Microcomputer-based, low cost method for measurement of spatial and temporal parameters of gait. *Journal of Biomedical Engineering*, **9**, 115–120.

Law H. T., Minns R. A. (1989). Measurement of the spatial and temporal parameters of gait. *Physiotherapy*, **75**, 81–84.

Lehmann J. F., Condon S. M., Price R., et al. (1987). Gait abnormalities in hemiplegia: their correction by ankle-foot orthoses. *Archives of Physical Medicine and Rehabilitation*, **68**, 763–771.

Lord M., Reynolds D. P., Hughes J. R. (1986). Foot pressure measurement: a review of clinical findings. *Journal of Biomedical Engineering*, **8**, 283–294.

Morris J. R. W. (1973). Accelerometry – a technique for the measurement of human body movements. *Journal of Biomechanics*, **6**, 729–736.

Muccio P., Andrews B., Marsolais E. B. (1989). Electronic orthoses: technology, prototypes, and practices. *Journal of Prosthetics and Orthotics*, **1**, 3–17.

Murray M. P. (1967). Gait as a total pattern of movement. *American Journal of Physical Medicine*, **46**, 290–333.

Murray M. P., Kory R. C., Clarkson B. H. (1969). Walking patterns in healthy old men. *Journal of Gerontology*, **24**, 169–178.

Murray M. P., Kory R. C., Sepic S. B. (1970). Walking patterns of normal women. *Archives of Physical Medicine and Rehabilitation*, **51**, 637–650.

Murray M. P., Sepic S. B., Gardner G. M., et al. (1978). Walking patterns of men with parkinsonism. *American Journal of Physical Medicine*, **57**, 278–294.

Murray M. P., Mollinger L. A., Sepic S. B., et al. (1983). Gait patterns in above–knee amputee patients: hydraulic swing control vs constant-friction knee components. *Archives of Physical Medicine and Rehabilitation*, **64**, 339–345.

New York University. (1986). *Lower Limb Orthotics*. New York: Prosthetics and Orthotics, New York University Postgraduate Medical School.

Olney S. J., Winter D. A. (1985). Selecting representative muscles for EMG analysis of gait: a methodology. *Physiotherapy Canada*, **37**, 211–217.

Paul J. P. (1965). Bio-engineering studies of the forces transmitted by joints. (II) Engineering analysis. In *Biomechanics and Related Bioengineering Topics*, (Kenedi J. P. ed.) Oxford: Pergamon, pp. 369–380.

Paul J. P. (1966). Forces transmitted by joints in the human body. *Proceedings of the Institute of Mechanical Engineers*, **181**, 8–15.

Perry J. (1969). The mechanics of walking in hemiplegia. *Clinical Orthopaedics and Related Research*, **63**, 23–31.

Perry J. (1974). Kinesiology of lower extremity bracing. *Clinical Orthopaedics and Related Research*, **102**, 18–31.

Perry J., Bekey G. A. (1981). EMG-force relationships in skeletal muscle. *CRC Critical Reviews in Biomedical Engineering*, **7**, 1–22.

Prodromos C. C., Andriacchi T. P., Galante J. O. (1985). A relationship between gait and clinical changes following high tibial osteotomy. *Journal of Bone and Joint Surgery*, **67A**, 1188–1194.

Robinson J. L., Smidt G. L. (1981). Quantitative gait evaluation in the clinic. *Physical Therapy*, **61**, 351–353.

Rose G. K. (1983). Clinical gait assessment: a personal view. *Journal of Medical Engineering and Technology*, **7**, 273–279.

Rose G. K. (1985). Use of ORLAU-Pedotti diagrams in clinical gait assessment. In *Biomechanical Measurement in Orthopaedic Practice* (Whittle M., Harris D., eds.) Oxford: Clarendon Press, pp. 205–210.

Saleh M., Murdoch G. (1985). In defence of gait analysis. *Journal of Bone and Joint Surgery*, **67B**, 237–241.

Saunders J. B. D. M., Inman V. T., Eberhart H. S. (1953). The major determinants in normal and pathological gait. *Journal of Bone and Joint Surgery*, **35A**, 543–558.

Shiavi R. (1985). Electromyographic patterns in adult locomotion: a comprehensive review. *Journal of Rehabilitation Research and Development*, **22**, 85–98.

Steindler A. (1953). A historical review of the studies and investigations made in relation to human gait. *Journal of Bone and Joint Surgery*, **35A**, 540–542.

Steven M. M., Capell H. A., Sturrock R. D, MacGregor J. (1983). The physiological cost of gait (PCG): a new technique for evaluating nonsteroidal antiinflammatory drugs in rheumatoid arthritis. *British Journal of Rheumatology*, **22**, 141–145.

Sutherland D. H., Cooper L. (1978). The pathomechanics of progressive crouch gait in spastic diplegia. *Orthopedic Clinics of North America*, **9**, 143–154.

Sutherland D. H., Olshen R. A., Biden E. N., Wyatt M. P. (1988). *The Development of Mature Walking*. London: Mac Keith Press.

Tashman S., Hicks R., Jendrzejczyk D. J. (1985). Evaluation of a prosthetic shank with variable inertial properties. *Clinical Prosthetics and Orthotics*, **9**, 23–28.

Todd F. N., Lamoreux L. W., Skinner S. R., et al. (1989). Variations in the gait of normal children. *Journal of Bone and Joint Surgery*, **71A**, 196–204.

Vaughan C. L., Murphy G. N., du Toit L. L. (1987). *Biomechanics of Human Gait: an Annotated Bibliography*. Champaign, Illinois: Human Kinetics Publishers.

Wall J. C., Charteris J., Turnbull G. I. (1987). Two steps equals one stride equals what?: the applicability of normal gait nomenclature to abnormal walking patterns. *Clinical Biomechanics*, **2**, 119–125.

Waters R. L., Garland D. E., Perry J., Habig T., Slabaugh P. (1979). Stiff–legged gait in hemiplegia: surgical correction. *Journal of Bone and Joint Surgery*, **61A**, 927–933.

Waters R. L., Lunsford B. R., Perry J., Byrd R. (1988). Energy–speed relationship of walking: standard tables. *Journal of Orthopaedic Research*, **6**, 215–222.

Whittle M. W. (1982). Calibration and performance of a three-dimensional television system for kinematic analysis. *Journal of Biomechanics*, **15**, 185–196.

Whittle M. W., Cochrane G. M., Chase A. P., et al. (1991). A comparison between two walking systems for paralysed people. *Paraplegia* (in press).

Winter D. A., Quanbury A. O., Reimer G. D. (1976). Analysis of instantaneous energy of normal gait. *Journal of Biomechanics*, **9**, 253–257.

Winter D. A. (1980). Overall principle of lower limb support during stance phase of gait. *Journal of Biomechanics*, **13**, 923–927.

Winter D. A. (1983). Energy generation and absorption at the ankle and knee during fast, natural, and slow cadences. *Clinical Orthopaedics and Related Research*, **175**, 147–154.

Winter D. A. (1985). Concerning the scientific basis for the diagnosis of pathological gait and for rehabilitation protocols. *Physiotherapy Canada*, **37**, 245–252.

Winter D. A. (1987). *The Biomechanics and Motor Control of Human Gait*. Waterloo, Ontario: University of Waterloo Press.

Winters T. F., Gage J. R., Hicks R. (1987). Gait patterns in spastic hemiplegia in children and young adults. *Journal of Bone and Joint Surgery*, **69A**, 437–441.

Index